The Best American Science and Nature Writing 2002

GUEST EDITORS OF
THE BEST AMERICAN SCIENCE
AND NATURE WRITING

2000 DAVID QUAMMEN

2001 EDWARD O. WILSON

2002 NATALIE ANGIER

The Best American Science and Nature Writing 2002

Edited and with an Introduction
by Natalie Angier

Tim Folger, Series Editor

HOUGHTON MIFFLIN COMPANY
BOSTON · NEW YORK 2002

Visit our Web site: www.houghtonmifflinbooks.com.

ISSN 1530-1508
ISBN 0-618-08297-2
ISBN 0-618-13478-6 (pbk.)

Printed in the United States of America

DOC 10 9 8 7 6 5 4 3 2 1

American Meritocracy, by Nicholas Lemann copyright © 1999 by Nicholas Lemann. Reprinted by permission of Farrar Straus and Giroux, LLC. SAT materials selected from 10 SATs. College Entrance Examination Board, 1983. Reprinted by permission of the College Entrance Examination Board, the copyright owner. Permission to reprint SAT materials does not constitute review or endorsement by Educational Testing Service or the College Board of this publication as a whole or of any questions or testing information it may contain.

"As Good as Dead" by Gary Greenberg. First published in *The New Yorker*, August 13, 2001. Copyright © 2001 by Gary Greenberg. Reprinted by permission of the author.

"Is That a Mountain Lion in Your Backyard?" by Gordon Grice. First published in *Discover*, June 2001. Copyright © 2001 by Gordon Grice. Reprinted by permission of Red Hourglass, Inc.

"The Dirt in the New Machine" by Blaine Harden. First published in the *New York Times Magazine*, August 12, 2001. Copyright © 2001 by Blaine Harden. Reprinted by permission of the author.

"Life's Rocky Start" by Robert M. Hazen. First published in *Scientific American*, April 2001. Copyright © 2001 by Scientific American, Inc. Reprinted by permission.

"Mothers and Others" by Sarah Blaffer Hrdy. First published in *Natural History*, May 2001. Copyright © 2002 by the University of Utah Press and the Trustees of the Tanner Lectures on Human Values. Reprinted from *The Tanner Lectures on Human Values*, Vol. 23, June 2002, as "The Past and Present of the Human Family" by permission of the University of Utah Press.

"Sound and Fury" by Garret Keizer. First published in *Harper's Magazine*, March 2001. Copyright © 2001 by Garret Keizer. Reprinted by permission of the author. "Bad Card" (Bob Marley) copyright © 1980 by Fifty-Six Hope Road Music Ltd./ Odnil Music Ltd./Blue Mountain Music Ltd. (PRS). All rights for the United States and Canada controlled and administered by Rykomusic, Inc. (ASCAP). International copyright secured. All rights reserved. Used by permission.

"The Pursuit of Innocence in the Golden State" by Verlyn Klinkenborg. First published in the *New York Times*, January 29, 2001. Copyright © 2001 by the *New York Times*. Reprinted by permission.

"Ripe for Controversy" by Robert Kunzig. First published in *Discover*, November 2001. Copyright © 2001 by Robert Kunzig. Reprinted by permission of the author.

"Wall Street Losses, Wall Street Gains" by Anne Matthews. First published in *Orion*, Spring 2001. Copyright © 2001 by Anne Matthews. Reprinted from *Wild Nights: Nature Returns to the City* by permission of North Point Press, a division of Farrar, Straus and Giroux, LLC.

"Dumb, Dumb, Duh Dumb" by Steve Mirsky. First published in *Scientific American*, November 2001. Copyright © 2001 by Scientific American, Inc. Reprinted by permission.

"I Have Seen Cancers Disappear" by Judith Newman. First published in *Discover*, May 2001. Copyright © 2001 by Judith Newman. Reprinted by permission of the author.

"How Islam Won, and Lost, the Lead in Science" by Dennis Overbye. First published in the *New York Times,* October 30, 2001. Copyright © 2001 by the *New York Times.* Reprinted by permission.

"A Little Reminder of Reality's Scale" by Chet Raymo. First published in the *Boston Globe,* July 24, 2001. Copyright © 2001 by Chet Raymo. Reprinted by permission of the author.

"Why McDonald's Fries Taste So Good" by Eric Schlosser. First published in *The Atlantic Monthly,* January 2001. Copyright © 2001 by Eric Schlosser. Reprinted from *Fast Food Nation* by permission of Houghton Mifflin Company.

"Shock and Disbelief" by Daniel Smith. First published in *The Atlantic Monthly,* February 2001. Copyright © 2001 by Daniel Smith. Reprinted by permission of the author. Excerpts from *A History of Psychology* by Edward Shorter copyright © 1997 by Edward Shorter. Reprinted by permission of John Wylie and Sons. Excerpt from *Toxic Psychiatry* by Peter Breggin copyright © 1991 by Peter Breggin. Reprinted by permission of St. Martin's Press, LLC.

"The Sting of the Assassin" by Peter Stark. First published in *Outside,* October 2001. Copyright © 2001 by Peter Stark. Reprinted from *Last Breath* by permission of Ballantine Books, a division of Random House, Inc.

"The Know-It-All Machine" by Clive Thompson. First published in *Lingua Franca,* September 2001. Copyright © 2001 by Clive Thompson. Reprinted by permission of the author.

"One Acre" by Joy Williams. First published in *Harper's Magazine,* 2001. Copyright © 2001 by Joy Williams. Reprinted by permission of International Creative Management.

"Very Dark Energy" by Karen Wright. First published in *Discover,* March 2001. Copyright © 2001 by Karen Wright. Reprinted by permission of the author.

Contents

Foreword xi

Introduction by Natalie Angier xv

ROY F. BAUMEISTER. *Violent Pride* 1
from *Scientific American*

BURKHARD BILGER. *Braised Shank of Free-Range Possum?* 10
from *Outside*

K. C. COLE. *Mind Over Matter* 21
from *The Los Angeles Times*

RICHARD CONNIFF and HARRY MARSHALL. *In the Realm of Virtual Reality* 24
from *Smithsonian*

FREDERICK C. CREWS. *Saving Us from Darwin* 34
from *The New York Review of Books*

BARBARA EHRENREICH. *Welcome to Cancerland* 58
from *Harper's Magazine*

H. BRUCE FRANKLIN. *The Most Important Fish in the Sea* 80
from *Discover*

MALCOLM GLADWELL. *Examined Life* 89
from *The New Yorker*

GARY GREENBERG. *As Good as Dead* 101
from *The New Yorker*

GORDON GRICE. *Is That a Mountain Lion in Your Backyard?* 114
from *Discover*

BLAINE HARDEN. *The Dirt in the New Machine* 124
from *The New York Times Magazine*

ROBERT M. HAZEN. *Life's Rocky Start* 137
from *Scientific American*

SARAH BLAFFER HRDY. *Mothers and Others* 148
from *Natural History*

GARRET KEIZER. *Sound and Fury* 161
from *Harper's Magazine*

VERLYN KLINKENBORG. *The Pursuit of Innocence in the Golden State* 179
from *The New York Times*

ROBERT KUNZIG. *Ripe for Controversy* 181
from *Discover*

ANNE MATTHEWS. *Wall Street Losses, Wall Street Gains* 185
from *Orion*

STEVE MIRSKY. *Dumb, Dumb, Duh Dumb* 196
from *Scientific American*

JUDITH NEWMAN. *"I Have Seen Cancers Disappear"* 198
from *Discover*

DENNIS OVERBYE. *How Islam Won, and Lost, the Lead in Science* 210
from *The New York Times*

CHET RAYMO. *A Little Reminder of Reality's Scale* 218
from *The Boston Globe*

ERIC SCHLOSSER. *Why McDonald's Fries Taste So Good* 221
from *The Atlantic Monthly*

DANIEL SMITH. *Shock and Disbelief* 234
from *The Atlantic Monthly*

PETER STARK. *The Sting of the Assassin* 255
from *Outside*

CLIVE THOMPSON. *The Know-It-All Machine* 266
from *Lingua Franca*

JOY WILLIAMS. *One Acre* 281
from *Harper's Magazine*

KAREN WRIGHT. *Very Dark Energy* 292
from *Discover*

Contributors' Notes 301

Other Notable Science and Nature Writing of 2001 306

Foreword

EVERY PROFESSION has its rite of passage, a crucible guaranteed to roil doubts and second thoughts about career choices. Pilots have their solo flights, surgeons their operations. For science journalists, it's that first crucial interview when they realize, with mounting unease, that they don't understand a single word of what some scientist is telling them. It happened to me several years ago. I had just started working as a reporter for *Discover* magazine and managed to convince my editor that I was ready to write a feature. One of the people I needed to interview for the story was an eminent physicist, a Nobel laureate. He graciously set aside two hours of his time one wintry afternoon in Princeton to talk to me about a perplexing problem in his field, a problem that was to be the subject of my article.

I turned on my tape recorder and asked my first question. In reply the physicist said something about an "antisymmetric total eigenfunction." It wasn't the sort of answer I was looking for. Worse, it wasn't the sort of answer I could understand. From there the gap between what the physicist said and what I followed could have been measured in megaparsecs. For the next 7,200 seconds I had almost no idea what this kindly, renowned, thoughtful gentleman was talking about. Sure, I could recognize the odd phrase here and there, but entire sentences might as well have been transmitted in a frequency range audible only to canines for all they meant to me. Somehow the few questions I sputtered during the remainder of the interview didn't betray my utter befuddlement and growing panic. For the most part I sat silently perspiring, nodding or grunting now and then to foster the illusion of comprehension.

When the interview finally ended I walked from the snow-covered campus to the train that would take me back to Manhattan, wondering how I would ever wring a story from such impenetrable raw material before my deadline. Over the next few weeks, after many more hours of interviews and phone conversations with perhaps a dozen physicists, I finished the assignment. The work was grueling, but satisfying.

That first interview turned out to be similar to many others in the years ahead. Although the panicky fear of failing to deliver a story eventually faded, the hard labor of translating the work of scientists into something that people will pay to read hasn't changed at all. Good writing is never easy, but writing about science is extraordinarily challenging. Most journalists, whether they're covering crime, politics, or business, can at least assume a common vocabulary, a certain degree of shared knowledge, on the part of their readers, not to mention their interview subjects. Science writers don't have that luxury. First they need to understand enough of the subject at hand to ask relevant questions. Then they must mold their interview notes and background reading of sundry science journals into a narrative that a reader will not just understand but enjoy. Not an easy profession.

Fortunately for us, there are many people who do it extremely well. The stories they tell are compelling, perhaps the most important of our time. J. Robert Oppenheimer, the controversial physicist who headed the Manhattan Project during World War II, once said, "Taken as a story of human achievement, and human blindness, the discoveries in the sciences are among the great epics."

The stories science tells us are not always comforting. Steven Weinberg, a Nobel laureate physicist (not the one who so confounded me years ago), has said that the more physicists study the universe, the more pointless it all seems. Scientists have not found any evidence of a special role for humanity in the scheme of things. Instead, human life looks like a very marginal phenomenon. Knowing that countless other species have arisen and disappeared on earth over the past 3 billion years, the existence of *Homo sapiens* seems less and less divinely ordained and ever more contingent.

When asked about Weinberg's bleak view, Jim Peebles, a Princeton astrophysicist, said, "I'm willing to believe that we are flotsam and jetsam."

But maybe those cold truths from the unflinching, vast perspective of science are what we need to hear. Genetic evidence suggests that every person now alive descends from the members of a small group of humans who lived in Africa between 100,000 and 200,000 years ago. Maybe the knowledge of our tentative, fragile place in the cosmos, and of our relatively recent common origin, marks the beginning of our maturity as a species. Maybe it's time to set aside the myths and legends that still sustain — and divide — so many of us.

Of course, such knowledge isn't welcome everywhere. As Frederick Crews writes in "Saving Us from Darwin," creationists still refuse to accept the full implications of *The Origin of Species* 143 years after its publication. They prefer to cling, using the most tortured reasoning, to a god who is "a glutton for praise," Crews writes. Their efforts to distort and suppress the teaching of science might seem ludicrous were they supported by only a few in our society. Unfortunately that's not the case, which makes Dennis Overbye's "How Islam Won, and Lost, the Lead in Science" disturbingly relevant. Science — and the liberal culture of tolerance and dispassionate inquiry that makes possible its pursuit — has many enemies. Perhaps the articles collected here will help win it a few more friends.

Working with Natalie Angier has been a particular pleasure for me — she was once my professor at New York University's graduate program in science writing. The only disadvantage of having Natalie as the guest editor is that none of her own writing could be included. Her reflections on the extraordinary sacrifices of New York's firemen, policemen, and other ordinary people on September 11 would have been one of my top picks for this volume. Search a library or the Internet for her story — "Altruism, Heroism, and Nature's Gifts in the Face of Terror," published in the September 18 edition of the *New York Times*.

I am very grateful to Deanne Urmy and Laura van Dam, my editors at Houghton Mifflin, for their good humor, guidance, and suggestions. Peter Brown, the former editor of *The Sciences,* put me in touch with Laura and Deanne. I keenly regret the demise of *The Sci-*

ences, one of the country's best magazines, which ceased publishing last year. Had it survived, I'm sure that it would have been represented in these pages. Burkhard Bilger, the editor of this series for the past two years and a contributor this year, offered much valuable advice. Finally, I can't adequately express my gratitude and love to Anne Nolan, who gave up Manhattan — and Brooklyn — to join me in Gallup, New Mexico.

TIM FOLGER

Introduction

In the immediate aftermath of September 11, when all light and sense, inflection and comprehension, seemed to vanish overmorning right along with those gorgeous, goofy, minimalist-maximalist twin towers, I was wracked with apocalyptic visions of a desolate world to come. The ancient curse of millennial psychosis had struck at last, I thought, and now my daughter would grow up in a time of brutal piousness, intolerance, and de-encephalization, as brigades of Truth Police roamed the streets, snarling Presa Canario dogs in tow.

So I wept and whined and flailed, and wrote violent little fantasy vignettes about mothers and daughters who figure out how to kill Osama bin Laden; and like so many people I couldn't sleep, night after night, and I exercised fiendishly, and drank a lot of white russians, and might really have fallen into a persistent state of vegetative gloom had I not started reading a book by Steven Weinberg, the Nobel laureate physicist and atheistic belletrist, called *Facing Up: Science and Its Cultural Adversaries.* I pounced on the chapter entitled "Before the Big Bang," in which Weinberg discusses various theories about the origins, or pre-origins, of the universe. To hypothesize about anything prior to the Big Bang, which brought time and physical laws as we know them into existence about 14 billion years ago, had long been taboo among astrophysicists: scientists don't like asking questions that they feel are impossible to answer, and this question seemed like the dooziest unsolvable problem of them all.

Happily, new theories such as superstrings and the inflationary model of the universe allow researchers to begin grappling with how the Bang came to be — or rather, the bang we know best. One version of inflation theory, called chaotic inflation, suggests an image of a nicely simmering pot of stew, with different bubbles of "scalar energy" popping up here and there, each a universe of its own. Very far away from our corner of the cosmos, Weinberg writes, "there may have been other big bangs before our own, and there may be others yet to come. Meanwhile the whole universe goes on expanding, so there is always plenty of room for more big bangs." He continued, "Thus although our own Big Bang had a definite beginning about 10 to 15 billion years ago, the bubbling up of new big bangs may have been going on forever in a universe that is infinitely old."

Certainly, this business of pre-bang astrophysics is very much a baby bubble of its own, and it may burst into nothingness, but Weinberg argues that the problem of origins is not beyond the reach of science. "We don't know if the universe is infinitely old or if there is a first moment," he concludes, "but neither view is absurd, and the choice between them will not be made by intuition, or by philosophy or theology, but by the ordinary methods of science."

Oh, how I loved that line and its almost coy blandness, and in that line I found my hope for the future in those moments when a new Age of Endarkenment leered large. I thought about the ferociously smart Weinberg, and his ambitious, hard-driving students, and the thousands of researchers and theoreticians like them at universities here and abroad, hammering away at "ordinary" science, which means solving equations of the sort of florid density so easily parodied in a *New Yorker* cartoon; and fighting for time at the Fermilab particle accelerator or on the Hubble space telescope; and crunching numbers and data on a mainframe for days or weeks at a time; and emerging at last with a reasonable portrait of places and spaces so unreasonably far and wide it makes my prolapsed mitral valve flutter just to think about them.

Equally astounding is what geologists have discovered about the history of our earth just by going to places such as the Grand Canyon and doing something more than complain about the tourists; or what paleontologists have learned about bestiaries gone by; or, more recently, what biologists have learned about the human genome — that it rambles, and repeats itself, and appears to be a La Brea tar pit for every drunken virus ever to stumble into an ancestor's cells.

In sum, when I felt really terrible, stuck in the deepest of tar-black funks, I thought about science and what it has wrought, and I knew that they will never win — *they* being the incuriosities of the world, whatever the particulars of their theology or ideology. They can never muzzle the mind, the many minds of science, the need to know. Science has been such a spectacular success — how could we ever give it up? It's like indoor plumbing: once you've tried it, the outhouse will never feel quite the same.

Beyond its extraordinary explanatory powers, science has another trait in its favor: its worldliness. If music is the universal language of the soul, science has become the lingua franca of the intellect. The greatest laboratories in this country are remarkably multicultural, the sort of places where you keep expecting Kofi Annan to walk in the door. Young researchers come to the United States from Europe, Asia, Africa, the Indian subcontinent, the Middle East, South America, Alpha Centauri, the Delta quadrant, even Canada. Once here (and please forgive this jingoistic spasm) in the world's greatest scientific candy store, they work together with the same goals and values: to do clever experiments well and elegantly, with the right controls, yielding results that any scientist, of any melanic, political, theistic, or zodiacal subtype, can recapitulate in the privacy of his or her own petri dish. In science, as in weapons negotiations, the code of conduct is "Trust, and verify." How nice it is not to have to take people at their dogma, but to be able to ask, in that snarly, whiny, chummy way that scientists do, what exactly the evidence is.

Yes, I love science, and I can't think of anything that is more worth writing about, or in greater need of good explanatory writing. For all their passion and productivity, scientists labor in profound anonymity. Last night, for example, while I was going over a list of some two dozen finalists for a big award given each year to a prominent woman in science, a friend stopped by to chat. I was curious to see whether any of the names on the list — many of them quite famous in their fields — would ring a bell with somebody who was neither a scientist nor a science journalist. As I ticked the names off one by one, my friend's head never stopped shaking. "Nope. Never heard of her. Nope, that one doesn't ring a bell either. Sorry, no."

"Okay, how about male scientists?" I asked. "Can you think of the name of any male scientist?"

"Louis Pasteur?" she offered.

"Well, it's a good thing scientists are such a cliquish lot," I said. "It's a good thing that most of them only care about winning the respect of their peers, because they sure aren't getting it from anybody else."

"What can you expect?" she said. "Most people don't understand the first thing about science, so they don't follow it. They don't have a clue."

This of course is an old refrain, one that I've had to cope with for the twenty-three years or so that I've been in the science writing business. People don't understand science. They're afraid of science. They're in awe of science. They're bored by science. They flunked high school chemistry. They're Barbie, they're Ken: "Math is hard!" And there's no denying it. Math is hard, science is hard, and it doesn't help that many high school chemistry teachers, if you threw them in the bay, would scare the sharks away.

But still. Is science really any harder than, say, Middle East politics? Or the fashion industry, for that matter? Look at a fashion magazine and explain to me who exactly wears those clothes, and what the obscure semiotics of the business are supposed to mean. My point being, science is a human endeavor like any other, and sure, it has its insiders who are possessive of their trade and expertise and use jargon like porcupines use their quills, smugly and defensively. Yet with a little effort, just about anybody can become reasonably literate in science, and it's well worth doing, and I'll even argue that it's natural to try. Think about it. Children love science and natural history museums, and they're often bored at art museums. Why should the opposite apply once a person becomes an adult? Why, for that matter, are science museums considered kiddie museums, when science is supposed to be so *hard*? Is it because we can see that children naturally want to understand how their world works, are insatiably, sometimes annoyingly curious? So what happens to drum that out of most people? Our highly imperfect educational system? Sure. But also, I believe, habit. It's become a bad, pervasive habit to think of science as a thing apart, an intellectual ghetto, so most people let their knees jerk autonomically. Science? Nope, never hear about it. I flunked high school chemistry, remember?

That's no excuse! I flunked my high school sewing class, but I still have a weakness for finely tailored clothes. Following science doesn't mean being able to practice science, any more than listening to music requires that you play an instrument or loving food requires you to study at the Culinary Institute of America. Science belongs to all of us, both metaphorically, because it's one of the great human enterprises, and literally, because we support it with our tax dollars. Perhaps if more people realized the degree to which we are the rightful owners of science, they would start feeling that highly effective emotion, a sense of entitlement, and would surmount their insecurities and replace them with that ultimate good, greed — the greedy, grasping need to know.

So where to go to know? Obviously there are hundreds, thousands of popular science books published every year, a significant fraction of them written by scientists. And though the platitude has it that most scientists are terrible writers, unable to hoist themselves out of obscurantism, pedantry, and the passive voice, in fact many are quite wonderful, and this has been true for a long time. Until a century or so ago, the line between science and literature was slim and porous. As Oliver Sacks points out in his memoirs, the great nineteenth-century chemist Humphry Davy published poetry as well as scientific papers: "his notebooks mix details of chemical experiments, poems, and philosophical reflections all together; and these did not seem to exist in separate compartments in his mind." Anton Chekhov, William Carlos Williams, Sir Arthur Conan Doyle — all practiced medicine, and it shows in their work. Albert Einstein wrote one of the best popular books about relativity theory, and Charles Darwin remains the finest explicator of Darwinism. One of my all-time favorite books, and the reason that I chose science writing as a career, was written by the physicist George Gamow in the 1940s. It is called *Mr. Tompkins in Wonderland,* and I urge you to read it; then you will understand as you never have before quantum physics, the curvature of space and time, the uncertainty principle, and why the best thing you can do in class is fall asleep.

More recently, scientist-authors have proliferated with yeastian speed, and again, many are wonderful and sometimes even best-selling authors: Weinberg, Sacks, Richard Dawkins, Lewis Thomas, Edward O. Wilson, Sarah Blaffer Hrdy, Alison Jolly, Frans de Waal, to name but a few. Yet, just as it doesn't take a doctorate in science or even a decent high school transcript to appreciate the beauty of science, so it is not the scientist alone who can write forcefully and accurately about science. Granted, I have no choice but to argue as much. I may have taken many science and math courses in college and afterward, and I've spent so much time in labs that I feel like a postdoc by proxy, but I don't have a science degree, and I wrote my senior thesis at Barnard College on William Faulkner's *Absalom, Absalom!*

However, I'm not merely being defensive here in saying that we benefit from the labors of good nonpedigreed science writers as much as we do from scientist-writers. For one thing, the list of blue-ribbon practitioners without portfolio is long, and includes Timothy Ferris, James Gleick, K. C. Cole, Elaine Morgan, John McPhee, Berton Roueché. In this volume, you'll find a piece by Frederick Crews that is the best critique I have ever read of the latest recrudescence of "creation science" known as intelligent design and of the larger, misbegotten movement to reconcile what is ultimately immiscible — science and religion. On the surface, "ID" theory is a bit more sophisticated than creationism. It doesn't try to find scientific evidence in support of a literal interpretation of the Bible; it accepts that the universe and the earth are billions rather than thousands of years old and that some evolution does occur. However, the specter of supernatural intervention remains; IDers have simply micrometized it, arguing that the cell is too complex and its parts too interdependent to have arisen by the mechanism of natural selection alone. And where a Darwinian watchmaker can't be found, they insist, a divine one must reside. In taking on this bafflingly popular movement, Crews presents the strengths of evolutionary theory with such sweeping meticulousness and confidence you'd think he'd been hammering natural selection theory or population genetics into students' heads for a generation. But no: he's an emeritus professor of English at the University of California.

Which brings me to the comparative advantages that non-scientist science writers occasionally may claim. Many biologists, when confronted with the fatuousness of creationism and its refurbished spores, refuse to engage in argument. Not only are they too busy doing legitimate research, they feel it is beneath them to take this stuff any more seriously than they would the astrology column of the local newspaper. Unfortunately, many Americans do not share scientists' curt dismissal of the antievolutionary line, and by resisting the call to debate, scientists have left the Darwin-knockers with all too much unopposed influence on school boards, among politicians, and in the writing of biology textbooks. Crews has nothing to lose by stepping in where evolutionary biologists sneer to tread, and in doing so he comes off heroically. I only hope his argument has impact where it counts: in every American classroom.

There's another thing that science writers will deign to do that many scientist-writers will not, one of particular relevance to this collection. They'll write articles for general-interest magazines and newspapers. Most scientists prefer to stick to books, perhaps because they don't want to waste time working on something that has such a short shelf life; writing for newspapers is particularly humbling, as one who has seen her articles being crumpled into fireplace fodder can attest. Moreover, general-interest articles are expected to incorporate viewpoints other than the scientist's own, and scientists may feel silly calling a rival for a quote. Yet articles obviously fill an essential niche. They're newsier than books, they're less opinionated, and they require far less commitment from the reader, if not always from the writer (meaty articles can take a year or longer to produce). And when an article is good, it is as sumptuous as any book and deserving of bookish preservation, which obviously is what we have in mind here.

In selecting articles, I had several criteria. I believe above all in clarity. If I reread a piece of mine some time after it was published and think, *Well, this part here is a little flaccid,* or *That line falls flat,* I may be disappointed; but if I come across something that is unclear, where even I, the writer, have no idea what the writer could have meant, I am mortified, furious with myself. Clarity is the foundation on which all else is built. As I used to tell students when I taught feature writing, you can disagree with my suggestions on how to fix a murky passage, but you can't talk me out of being confused; the passage simply isn't well enough to leave alone. Clarity of expression is particularly important when describing a tough bit of science. I'm grateful to Karen Wright for helping me to imagine a nearly unfathomable form of energy — the "dark energy" that seems to push things apart and may explain why the rate of expansion of our universe keeps getting faster and faster. Likewise, Robert M. Hazen's description of the role of minerals in the origins of life is stylistically as well as substantively crystalline. I am also indebted to Gary Greenberg for laying out with such eloquence a technically, politically, and emotionally demanding issue like what it means to say that a patient who breathes and whose heart still beats is nonetheless "brain dead" and thus is ripe for organ picking.

Good writing is clear, and it is interesting to read. That may sound obvious, but I have yet to figure out what exactly makes a piece interesting. You need cadence, agile verbs, exacting yet surprising metaphors, yes yes yes. But more than that, I think, good writing, like good gnosis, is the product of greed, of writers who greedily claim their subjects for themselves and absorb every detail, until the story is part of the latticework of their cells. Then and only then can they write with the passion of converts, of those who believe that this story is *the* most interesting story they have ever heard.

It helps when the person has lived the story in question. Barbara Ehrenreich is free to express her aversion to the pink-ribboned Breast Cancer Awareness movement because she has had breast cancer, and she doesn't want anybody to try smothering her outrage by plying her with Wish Upon a Star teddy bears, boxes of crayons, and invitations to industry-sponsored 6K races or gushing about how breast cancer makes one a "better person" who appreciates the really "important" things in life and who, in the wake of chemotherapy, may even have smoother skin and chirpier hair! Joy Williams makes the case for Nature Unplugged by telling the story of her one acre in Florida and how she let it grow as weedily wild as it pleased, whether it pleased her neighbors or no.

Other stories, while less immediately first-person, nonetheless retain the urgency of the personal. In Anne Matthews's gorgeous story about the ecology and biogeography of Manhattan, you can almost feel the still-warm bodies of songbirds that litter the streets of Wall Street before dawn, having been mesmerized by the lights of the skyscrapers into flying round and round in circles and finally falling, exhausted, to their death. In a poignant touch, much of the story's action takes place at the base of the then-lofty World Trade Center. As I read Gordon Grice's story about the recent incursion of mountain lions into the suburbs of western states, I had a powerful upwelling of contradictory emotions. *Hats off to the cats that came back!* squealed my eco-conscience, my conviction that all creatures have a right to persist, German cockroaches excepted. But oh, lord, what a horror it was for Barbara Schoener, a long-distance runner, who in 1994 was attacked by a mountain lion on a trail in El Dorado County. The evidence from the torn-up embankment suggested that Schoener put up a good fight, but the cougar killed her in the end. Just how much nature are any of us willing to nurture when our own necks, or the necks of our children, are at stake?

Another approach to telling a great story is to take chances and to be willing to risk looking foolish. I admire Peter Stark for having taken just that gamble in "The Sting of the Assassin," in which he uses a fictional device — the story of a couple on a less than starry-eyed honeymoon — to describe in exquisite detail how venom affects the human body. The style is arch, almost campy, but it works much better than I might have predicted, and it makes me grumpy that I didn't think of the technique myself.

As it happens, a number of the pieces are about food or the food chain: why McDonald's french fries taste so good; why French cheese won't taste good if you start with pasteurized milk; why Westerners happily eat catfish, crawfish, snails, and frog legs but continue to balk at braised possum, armadillo cheeks, and fried mink; and why the most important food fish in the sea is a fish we don't eat, the oily, foul, kipper-like fish called menhaden. And then there is a story about the ones who feed us first and best: mothers.

In the last quarter century, science writing has changed substantially. It is less gee-whizzy and gullible than it used to be, more surefooted. The article about Steven Rosenberg, a renowned tumor biologist, exemplifies this trend. He is not hyped and lionized as the man with the next big answer to cancer, as he might have been in the past. Instead, he is presented as somebody thrashing about in the trenches, trying whatever he can to help patients who are almost beyond salvation, exulting over the occasional victory, admitting to the far more frequent failures. Rosenberg's story is the story of contemporary cancer research: enormous progress on the basic research front and very little in applying that knowledge to the bedside.

Perhaps the clearest sign that science writing has matured and is seated comfortably at the literary dining table is the impressive array of science essayists out there, writers who can convey complex ideas in a few deft, plangent paragraphs. The best essayists appeal simultaneously to the cognitive and emotional domains of the brain, the Apollo and Dionysus within, so that you feel you have learned something and fallen in love all at once.

Read, think, and be merry. The universe is expanding. May our minds follow suit.

NATALIE ANGIER

ROY F. BAUMEISTER

Violent Pride

FROM *Scientific American*

SEVERAL YEARS AGO a youth counselor told me about the dilemma he faced when dealing with violent young men. His direct impressions simply didn't match what he had been taught. He saw his violent clients as egotists with a grandiose sense of personal superiority and entitlement, but his textbooks told him that these young toughs actually suffered from low self-esteem. He and his staff decided they couldn't go against decades of research, regardless of what they had observed, and so they tried their best to boost the young men's opinions of themselves, even though this produced no discernible reduction in their antisocial tendencies.

The view that aggression stems from low self-esteem has long been common knowledge. Counselors, social workers, and teachers all over the country have been persuaded that improving the self-esteem of young people is the key to curbing violent behavior and to encouraging social and academic success. Many schools have students make lists of reasons why they are wonderful people or sing songs of self-celebration. Many parents and teachers are afraid to criticize kids, lest it cause serious psychological damage and turn some promising youngster into a dangerous thug or pathetic loser. In some sports leagues, everyone gets a trophy.

A number of people have questioned whether these feel-good exercises are really the best way to build self-esteem. But what about the underlying assumption? When my colleagues and I began looking into the matter in the early 1990s, we found article after article citing the "well-known fact" that low self-esteem causes violence. Yet we were unable to find any book or paper that offered a formal statement of that theory, let alone empirical evidence to support it. Everybody knew it, but nobody had ever proved it.

Unfortunately for the low self-esteem theory, researchers have gradually built up a composite image of what it is like to have low self-esteem, and that image does not mesh well with what we know about aggressive perpetrators. People who have a negative view of themselves are typically muddling through life, trying to avoid embarrassment, giving no sign of a desperate need to prove their superiority. Aggressive attack is risky; people with low self-esteem tend to avoid risks. When people with low self-esteem fail, they usually blame themselves, not others.

Faced with these incongruities, we cast about for an alternative theory. A crucial influence on our thinking was the seemingly lofty self-regard of prominent violent people. Saddam Hussein is not known as a modest, cautious, self-doubting individual. Adolf Hitler's exaltation of the "master race" was hardly a slogan of low self-esteem. These examples suggest that self-esteem is indeed an important cause of aggression — high, that is, not low self-esteem.

We eventually formulated our hypothesis in terms of threatened egotism. Not all people who think highly of themselves are prone to violence. That favorable opinion must be combined with some external threat to the opinion. Somebody must question it, dispute it, undermine it. People like to think well of themselves, and so they are loath to make downward revisions in their self-esteem. When someone suggests such a revision, many individuals — those with inflated, tenuous, and unstable forms of high self-esteem — prefer to shoot the messenger.

It would be foolish to assert that aggression always stems from threatened egotism or that threatened egotism always results in aggression. Human behavior is caused and shaped by various factors. Plenty of aggression has little or nothing to do with how people evaluate themselves. But if our hypothesis is right, inflated self-esteem increases the odds of aggression substantially. For those aggressive acts that do involve the perpetrators' self-regard, we believe that threatened egotism is crucial. Obviously, this new theory could have implications for designing effective methods to reduce violence.

So how does a social psychologist establish whether low or high self-esteem leads to violence? Because there is no perfect, general method for understanding complex questions about human beings, social scientists typically operate by conducting multiple studies with different methods. A single study can be challenged, especially if competing views exist. But when a consistent pattern emerges, the conclusions become hard to ignore.

Researchers measure self-esteem by asking a standardized series of questions, such as "How well do you get along with other people?" and "Are you generally successful in your work or studies?" The individual chooses from a range of responses, and the overall score falls somewhere on the continuum from negative to positive. Strictly speaking, it is misleading to talk of "people high in self-esteem" as if they were a distinct type, but the need for efficient communication pushes researchers into using such terms. By "people high in self-esteem," I refer broadly to those who scored above the median on the self-esteem scale. Statistical analyses respect the full continuum.

Many laypeople have the impression that self-esteem fluctuates widely, but in fact these scores are quite stable. Day-to-day changes tend to be small, and even after a serious blow or boost, a person's self-esteem score returns to its previous level within a relatively short time. Large changes most often occur after major life transitions, such as when a high school athlete moves on to college to find the competition much tougher.

Quantifying aggression is trickier, but one approach is simply to ask people whether they are prone to angry outbursts and conflicts. These self-reported hostile tendencies can then be compared to the self-esteem scores. Most research has found a weak or negligible correlation, although an important exception is the work done in the late 1980s by Michael H. Kernis of the University of Georgia and his colleagues. They distinguished between stable and unstable self-esteem by measuring each person's self-esteem on several occasions and looking for fluctuations. The greatest hostility was reported by people with high but unstable self-esteem. Individuals with high, stable self-esteem were the least hostile, and those with low self-esteem (either stable or unstable) were in between.

Another approach is to compare large categories of people. Men on average have higher self-esteem than women and are also more aggressive. Depressed people have lower self-esteem and are less violent than nondepressed people. Psychopaths are exceptionally prone to aggressive and criminal conduct, and they have very favorable opinions of themselves.

Evidence about the self-images of specific murderers, rapists, and other criminals tends to be more anecdotal than systematic, but the pattern is clear. Violent criminals often describe themselves as superior to others — as special, elite persons who deserve preferential treatment. Many murders and assaults are committed in response to blows to self-esteem such as insults, "dissing," and humiliation. (To be sure, some perpetrators live in settings where insults threaten more than their opinions of themselves. Esteem and respect are linked to status in the social hierarchy, and to put someone down can have tangible and even life-threatening consequences.)

The same conclusion has emerged from studies of other categories of violent people. Street-gang members have been reported to hold favorable opinions of themselves and to turn violent when these views are disputed. Playground bullies regard themselves as superior to other children; low self-esteem is found among the victims of bullies but not among bullies themselves. Violent groups generally have overt belief systems that emphasize their superiority over others. War is most common among proud nations that feel they are not getting the respect they deserve, as Daniel Chirot discusses in his fascinating book *Modern Tyrants.*

Drunk people are another such category. It is well known that alcohol plays a role in either a majority or a very large minority of violent crimes; booze makes people respond to provocations more vehemently. Far less research has examined the link with self-esteem, but the findings do fit the egotism pattern: consuming alcohol tends to boost people's favorable opinions of themselves. Of course, alcohol has myriad effects, such as impairing self-control, and it is hard to know which is the biggest factor in drunken rampages.

Aggression toward the self exists, too. A form of threatened egotism seems to be a factor in many suicides. The rich, successful person who commits suicide when faced with bankruptcy, disgrace, or scandal is an example. The old, glamorous self-concept is no longer tenable, and the person cannot accept the new, less appealing identity.

*

Taken together, these findings suggest that the low self-esteem theory is wrong. But none involves what social psychologists regard as the most convincing form of evidence: controlled laboratory experiments. When we conducted our initial review of the literature, we uncovered no lab studies that probed the link between self-esteem and aggression. Our next step, therefore, was to conduct some. Brad J. Bushman of Iowa State University took the lead.

The first challenge was to obtain reliable data on the self-concepts of participants. We used two different measures of self-esteem, so that if we failed to find anything, we could have some confidence that the null result was not simply an artifact of having a peculiar scale. Yet we were also skeptical of studying self-esteem alone. The hypothesis of threatened egotism suggested that aggressive behavior would tend to occur among only a subset of people with high self-esteem. In the hope of identifying this subset, we tested for narcissism.

As defined by clinical psychologists, narcissism is a mental illness characterized by inflated or grandiose views of self, the quest for excessive admiration, an unreasonable or exaggerated sense of entitlement, a lack of empathy (that is, being unable to identify with the feelings of others), an exploitative attitude toward others, a proneness to envy or wish to be envied, frequent fantasies of greatness, and arrogance. The construct has been extended beyond the realm of mental illness by Robert Raskin of the Tulsa Institute of Behavioral Sciences in Oklahoma and several of his colleagues, who have constructed a scale for measuring narcissistic tendencies.

We included that measure alongside the self-esteem scales, because the two traits are not the same, although they are correlated. Individuals with high self-esteem need not be narcissistic. They can be good at things and recognize that fact without being conceited or regarding themselves as superior beings. The converse — high narcissism but low self-esteem — is quite rare, however.

The next problem was how to measure aggression in the laboratory. The procedure we favored involved having pairs of volunteers deliver blasts of loud noise to each other. The noise is unpleasant and people wish to avoid it, so it provides a good analogue to physical aggression. The famous social psychology experiments of the 1960s used electric shock, but safety concerns have largely removed that as an option.

The noise was presented as part of a competition. Each participant vied with somebody else in a test of reaction time. Whoever responded more slowly received a blast of noise, with the volume and duration of the noise set by his or her opponent. This procedure differed from that of earlier studies, in which the subject played the role of a "teacher" who administered noise or shock to a "learner" whenever the learner made a mistake. Critics had suggested that such a method would yield ambiguous results, because a teacher might deliver strong shocks or loud noise out of a sincere belief that it was an effective way to teach.

To study the "threat" part of threatened egotism, we asked participants to write a brief essay expressing their opinion on abortion. We collected the essays and then (ostensibly) redistributed them, so the two contestants could evaluate each other's work. Each participant then received his or her own essay back with the comments and ratings that the other person had (supposedly) given it.

In reality, the essays that people graded were fakes. We took the real essays aside and randomly marked them good or bad. The good evaluation included very positive ratings on all counts and the handwritten comment, "No suggestions, great essay!" The bad evaluation contained low marks and the additional comment, "This is one of the worst essays I have read!" After handing back the essays and evaluations, we gave out instructions for the reaction-time test and the subjects began to compete. The measure of aggression was the level of noise with which they blasted each other.

The results supported the threatened-egotism theory rather than the low self-esteem theory. Aggression was highest among narcissists who had received the insulting criticism. Non-narcissists (with either high or low self-esteem) were significantly less aggressive, as were narcissists who had been praised.

In a second study, we replicated these findings and added a new twist. Some participants were told that they would be playing the reaction-time game against a new person — someone different from the person who had praised or insulted them. We were curious about displaced aggression: would people angered by their evaluation lash out at just anybody? As it happened, no. Narcissists blasted people who had insulted them but did not attack an innocent third party. This result agrees with a large body of evidence that violence against innocent bystanders is, despite conventional wisdom, quite rare.

A revealing incident illuminates the attitudes of the narcissists. When a television news program did a feature on this experiment, we administered the test to several additional participants for the benefit of the cameras. One of them scored in the 98th percentile on narcissism and was quite aggressive during the study. Afterward he was shown the film and given the opportunity to refuse to let it be aired. He said to put it on — he thought he looked great. Bushman then took him aside and explained that he might not want to be seen by a national audience as a highly aggressive narcissist. After all, the footage showed him using severe profanity when receiving his evaluation, then laughing while administering the highest permitted levels of aggression. The man shrugged this off with a smile and said he wanted to be on television. When Bushman proposed that the television station at least digitize his face to disguise his identity, the man responded with an incredulous no. In fact, he said, he wished the program could include his name and phone number.

Would our laboratory findings correspond to the outside world? Real-life violent offenders are not the easiest group of people to study, but we gained access to two sets of violent criminals in prison and gave them the self-esteem and narcissism questionnaires. When we compared the convicts' self-esteem with published norms for young adult men (mostly college students) from two dozen different studies, the prisoners were about in the middle. On narcissism, however, the violent prisoners had a higher mean score than any other published sample. It was the crucial trait that distinguished these prisoners from college students. If prison seeks to deflate young men's delusions that they are God's gift to the world, it fails.

A common question in response to these findings is, "Maybe violent people seem on the surface to have a high opinion of themselves, but isn't this just an act? Mightn't they really have low self-esteem on the inside, even if they won't admit it?" This argument has a logical flaw, however. We know from ample research that people with overt low self-esteem are not aggressive. Why should low self-esteem cause aggression only when it is hidden and not when it is overt? The only difference between hidden and overt low self-esteem is the fact of its being hidden, and if that is the crucial difference, then the cause of violence is not the low self-esteem itself but the concealment of it. And what is concealing it is the veneer of egotism — which brings us back to the threatened egotism theory.

Various researchers have tried and failed to find any sign of a soft inner core among violent people. Martin Sanchez-Jankowski, who spent ten years living with various gangs and wrote one of the most thorough studies of youth gang life, had this to say: "There have been some studies of gangs that suggest that many gang members have tough exteriors but are insecure on the inside. This is a mistaken observation." Dan Olweus of the University of Bergen in Norway has devoted his career to studying childhood bullies, and he agrees: "In contrast to a fairly common assumption among psychologists and psychiatrists, we have found no indicators that the aggressive bullies (boys) are anxious and insecure under a tough surface."

The case should not be overstated. Psychology is not yet adept at measuring hidden aspects of personality, especially ones that a person may not be willing to admit even to himself or herself. But at present there is no empirical evidence or theoretical reason that aggressors have a hidden core of self-doubt.

Although this conclusion contradicts the traditional focus on low self-esteem, it does not mean that aggression follows directly from an inflated view of self. Narcissists are no more aggressive than anyone else, as long as no one insults or criticizes them. But when they receive an insult — which could be a seemingly minor remark or act that would not bother other people — the response tends to be much more aggressive than normal. Thus, the formula of threatened egotism combines something about the person with something about the situation. Whatever the details of cause and effect, this appears to be the most accurate formula for predicting violence and aggression.

These patterns raise misgivings about how schools and other groups seek to boost self-esteem with feel-good exercises. A favorable opinion of self can put a person on a hair trigger, especially when this favorable opinion is unwarranted. In my view, there is nothing wrong with helping students and others to take pride in accomplishments and good deeds. But there is plenty of reason to worry about encouraging people to think highly of themselves when they haven't earned it. Praise should be tied to performance (including improvement) rather than dispensed freely as if everyone had a right to it simply for being oneself.

The person with low self-esteem emerges from our investigation as someone who is not prone to aggressive responses. Instead one should beware of people who regard themselves as superior to others, especially when those beliefs are inflated, weakly grounded in reality, or heavily dependent on having others confirm them frequently. Conceited, self-important individuals turn nasty toward those who puncture their bubbles of self-love.

BURKHARD BILGER

Braised Shank of Free-Range Possum?

FROM *Outside*

"WHAT WE HAVE HERE is a radial pattern of wild meats," Jeff Jackson says, pointing his spatula at a cast-iron skillet. Four small mounds of mangled protein, each a different shade and texture, lie in a perfect parabola, like tissue samples from a crime lab. "First you'll eat them," Jackson says. "Then I'll tell you what they are."

Lifting my fork, I probe a mushroom cap brimming with a gray, speckled, liverish substance. To my right, Jackson's wife, Phyllis, picks at her salad and watches. "Back before we were married, we spent a whole summer living off roadkill," she says. "I remember one time, we ate a mink. That was one tough little animal. Can't say I liked the taste, either. There was this urine flavor, like the kidneys hadn't filtered out all the impurities." Jeff settles into the chair across from me. "Leeches were disappointing, too," he sighs. "Tasted just like the marinade. Didn't have any leech flavor *at all.*"

Glancing up at their expectant faces, I feel a wave of peer pressure such as I haven't experienced since junior high. It's early April, and the air is thick with the scent of sweet gums and pines, of things busy being born and busy dying. I have come to Georgia to expand my palate, to see what pockets of resistance remain in the South to the advancing army of Whoppers and Big Macs. But I was hoping to ease into the topic more gradually. The Jacksons, I thought, could offer a sober, academic accounting of the politics and economics of hunting and gathering for one's own table. After all, Jeff, sixty, is a professor of wildlife management in the School of

Forest Resources at the University of Georgia, and Phyllis, fifty-five, rounds out her homemaking and carpentry by documenting the vegetation of the Smoky Mountains for the University of Georgia's Center for Remote Sensing and Mapping. But scientific dispassion, I find, makes its own gustatory demands.

Two centuries ago, an explorer would have thought nothing of sitting down at a stranger's table and eating whatever flesh was placed before him. When Indian guides led British explorer John Lawson through the Carolina wilderness in 1701, he dined on beaver, polecat, and bear, among other delicacies. ("A roasted or barbakued Turkey, eaten with Bears Fat, is held a good Dish," Lawson wrote in his diary. "And indeed, I approve of it very well; for the Bears Grease is the sweetest and least offensive to the Stomach . . . of any Fat of Animals I ever tasted.") As late as 1909, the Atlanta Chamber of Commerce served persimmon beer, turtle soup, and barbecued possum to the president-elect. "Surely the famous smile of William Howard Taft never kindled across a happier evening," a reporter for the *Atlanta Journal-Constitution* wrote. Years later, another reporter at the paper mused: "It was believed to be the last time that a U.S. president supped on marsupial."

But in the decades since, the American diet has cut loose from its wilderness moorings. I grew up in Oklahoma, where southern cooking once left off and cowboy carbo-loading began. Yet after two years spent researching and writing a book on clandestine southern traditions, I am led to believe that everyone eats the same stuff. To most northerners, the South is the last refuge of strange food — of Moon Pies and pig organs and pickled eggs bobbing pinkly next to the cash register. But from what I've seen, southerners are the least adventurous eaters of all. Their cities are girdled with an extra layer of fast food, their vegetables invariably canned or overcooked, their palates tuned to the twin wavelengths of ketchup and processed meat. And so I've left the beltways and strip malls and gone in search of something more savory.

Now, chewing on another rich yet fibrous flap of mystery muscle, I wonder what it is, exactly, that makes something inedible. Is it just a matter of physiology, of nutrient deficiencies and taste-bud densities, or is it more psychological — a habit of mind shaped by culture, temperament, and parents telling us to eat our vegetables? Taste is a mind-body problem of the most intractable kind, and nothing brings it into focus as vividly as eating something unknown and potentially disgusting. Jeff, smiling faintly, informs me that this particular mouthful is armadillo meat. Why should that make my throat constrict and my stomach leap into my diaphragm? Does the fact that some southerners call this "possum on the half shell" make it any less palatable? The answers may determine whether true southern food can ever rise again.

The Jacksons, I'm happy to report, no longer content themselves with roadkill — though Jeff says he might eat a monkey if it was served to him. In fact, they exemplify a new kind of southern land ethic, one that is cosmopolitan yet self-sufficient, discerning yet omnivorous. In their house, everything has a dual purpose: the chimney is a nest for swifts; the cabinets, made of salvaged oak, are a lesson in recycling; the pear trees are food for deer and an orchard of heirloom species. Whenever a hunter leases his land, Jeff takes him around front to see the pyramid of deer skulls nailed to the front of the house. Arranged in order of size, with the largest on top, the skulls are a point of pride — an austere decoration, a warning to trespassers, and above all, a teaching aid. *See those?* Jeff says to the hunters, pointing to the ones with horns just budding from their foreheads. *Those are less than a year old. Don't kill those.*

One afternoon, Jeff takes me on a tour of his 350 acres, a patchwork of hardwood forest, hay meadow, and fruit trees outside the town of Arnoldsville. With his graying beard and kindly manner, his beat-up hat and blue eyes that go wide with feigned amazement, he looks like a latter-day Merlin. Living off the land isn't worth the bother anymore, he declares — "In terms of protein per effort, you can't justify it in any way" — but hunting still has its rewards.

Jackson believes hunting is the missing link in most Americans' environmental education. Boys once learned how a forest works from spending hours in it, keeping perfectly still. But from 1975 to 1996, according to the U.S. Fish and Wildlife Service, annual sales in hunting licenses dropped from 16.5 million to 15.2 million, and the percentage of Americans who hunt dropped even further. To Jeff, the main effect has been a rise not so much in ignorance as in sentimentality.

Take our attitudes toward deer, he says. As hunters have dwindled, deer have multiplied: there are nearly as many whitetails now as when the Pilgrims arrived, though only 4 percent of America's old-growth forests are still standing. As a result, songbirds and small mammals are being browsed out of house and home, rare plants are under attack, and hundreds of thousands of motorists crash into deer every year. And yet when new hunting permits are granted in places like Hilton Head, South Carolina, to try to control deer, wildlife-conservation groups protest and file lawsuits. "Deer aren't stupid," Jeff insists. "They realize that the rules have changed. Subdivisions used to be where they got shot; now it's where they're safe." As for the Jacksons, 95 percent of the red meat they eat is venison. Whenever the freezer is empty, Jeff simply wanders out into his meadow with a shotgun. "If I see a deer, I'll shoot it," he says.

The next morning Jeff has a class to teach in Athens, and I have hundreds of miles to drive by noon. But we head into the woods before dawn anyway, to stalk wild turkeys. Hunters sometimes go an entire season without bagging a bird. Nevertheless, only minutes from the house Jeff signals for me to stop. "Hear that?" he whispers, as a chortling sound echoes through the oaks. "That's the love song of the male turkey." He sits down, pulls a small cedar box from his pocket, and gently draws the lid across the frame, mimicking the female turkey's high, piping response. A few calls and responses later, he lifts his 12-gauge double-barrel and sends a ragged blast ripping through the trees. "This was an efficient hunt," he says, carrying the turkey back to the house by its feet. "It didn't take time away from other income-generating activities." He wraps the bird in a garbage bag, throws it into the trunk of his car, and strips off his camouflage. Beneath his canvas coveralls, a suit and tie emerge perfectly clean and unwrinkled, ready for his morning class.

As I head south from the Jacksons' across the Piedmont hills, the country clubs and subdivisions give way to brick-and-magnolia county seats and the tarmac turns to red clay. "This program is brought to you by the Last Resort Grill," an Atlanta station announces, "featuring nouvelle southern cuisine in a casually elegant setting enhanced by the work of local artists."

Even as hunters become an endangered species, more people are choosing to have their wilderness served to them on a platter.

Not long ago, catfish were considered fit only for other bottom-feeders; now U.S. farms grow more than half a billion pounds every year. Crawfish gross on average $50 million annually in Louisiana alone, and other animals are making the same transition. In the last fifteen years, bison ranches, deer ranches, pigeon, alligator, and turtle farms have sprung up across the South, and urbanites have begun to develop a taste for game. In Boston, at Savenor's Market, kangaroo meat sells for $15.99 a pound, camel for $39.99, lion for $19.99, and zebra for $39.99. All of it is raised on game farms in the United States.

Over the next few days I meander through Georgia's sand hills and along the coastal plain, gathering wild meats as I go. In Alapaha I talk to a man named Ken Holyoak who traps turtles for the voodoo trade and other religious purposes. "The Haitians, they put the turtle on a pedestal and worship him," he says. "The Cubans tie him to their stomach and dance around." He also claims to have perfected the first system for mass-producing bullfrogs. (I leave with one of each.) I investigate the cult of the wild ramp — "the world's most potent onion" — and watch a young man called Big Foot shoot, skin, and gut a wild hog. (He gives me the "back strap" to consume later.) But to find a true test for the modern American palate, I have to cross the border into Alabama, favored dwelling place of the creature once savored by William Howard Taft.

Opossums are America's great "underutilized meat," Jeff Jackson told me: plentiful, easy to catch, and twice as high in protein as beef cattle, pound for pound. Opossums were here before the Ice Age and will likely be here long after global warming. Like frogs and squirrels, they were a subsistence food for generations, their fat a godsend to calorie-starved settlers. For years, Appalachian families fattened one up every fall for Thanksgiving dinner. But when packaged foods came in and fatty foods went out, marsupials were the first items dropped from the menu. What culinary currency the animal still has is due almost entirely to one man: Frank Basil Clark.

Born in the North Carolina hills in 1930 and raised on anything he could kill, Clark moved to Clanton, Alabama, as a young man and managed a drive-in theater. It was there, in 1969, that a thought struck him with the force of revelation: "America has put a man on the moon, but it hasn't done a thing with the possum."

Clark was living in a mobile home at the time, and he had no real experience as a revolutionary, but he knew this thing was bigger than just one man. Pooling his meager resources, he founded the Possum Growers & Breeders Association of America in 1971 and hatched plans to breed a superpossum. He organized possum beauty contests and crowned a Possum Queen. He convinced the U.S. Agency for International Development to look into raising possums for food and physically handed a particularly handsome specimen to Richard Nixon during a presidential campaign stop in Birmingham in 1972. Most of all, he flooded the country with his bumper stickers, turning his motto into a redneck rallying cry: "Eat More Possum." The possum campaign grew beyond any rational bounds. By the mid-1980s the PGBAA's membership had ballooned to more than 100,000, and Clanton was an obligatory photo-op for presidential candidates. "George Bush Sr. said if he was elected, me and a possum could spend the night in Abraham Lincoln's bed," Clark remembers.

Unfortunately, by the time I arrive in Clanton, the possum madness has died down. Buoyed by his celebrity, Clark served two terms as mayor, from 1976 to 1984, but realpolitik seems to have let the air out of his possum propaganda. In the late eighties he took a job installing telephone lines and left town for a spell. He's back now, still spinning the same spiel, but PGBAA membership is way down, and even the bumper stickers have started to grow scarce. "I lost a bunch of good possum growers to them crazies," Clark says, referring to activists with People for the Ethical Treatment of Animals, who "wouldn't leave [him] alone" for a while. "You're not one of them crazies, are you?"

Bulb-nosed and rheumy-eyed, with a widening gut, Clark proudly shows me the swimming pool he had built as mayor, and the nice new curbs in the black part of town. But when I ask him for some possum meat for the road, he looks away uncomfortably. "If I start sellin' possums," he says, "the other growers will think the market is bad." And that would presumably lead to a frenzied sell-off — the possum world's own Black Tuesday. Eventually we have to drive out to his friend Barney's house to find an animal. "I don't like to eat 'em much," Barney says, dropping the writhing animal in Clark's cage. "I just like to catch 'em in the woods and then watch 'em with their babies."

These days, Clark's son Tom does most of the possum breeding in town — though he's shifted his sights from feeding the poor to starting a possum theme park. Not long ago, a church in Opelika called to see if he might bring a couple of possums to a wild-game cookout they were hosting. There would be a thousand guests or more, they told him, and crow and squirrel would be on the menu as well. "I picked some nice ones ahead of time and fattened 'em up good," Tom says. "But when I called the preacher later, he said people hardly touched 'em. 'We had a lot of folks,' he told me, 'but we had too much possum.'"

What happened? Why can Americans stomach catfish and crawfish — not to mention scrapple and Twinkies and electric-blue Gatorade — but not possum?

Thirty years ago, the anthropologist Claude Levi-Strauss divided all food into a "culinary triangle" — the raw, the cooked, and the rotten — and then examined how cultures deal with each category. The French have little tolerance for raw food, he wrote, while Italians love it. Americans can't stand rot — after D-day, GIs sometimes destroyed Normandy dairies because their cheeses smelled like decaying corpses — whereas some American Indians preferred a rotted buffalo carcass to a freshly killed one. "The cooking of a society," Levi-Strauss concluded, "is a language in which it unconsciously translates its structure."

The most famous case of dietary prejudice, if only because so many cultures have had to grapple with it, is in the Bible. "These shall ye not eat," Leviticus declares, and proceeds to list half the animals of ancient Israel: camels, coneys, hares, and swine; fish without fins or scales; eagles, ospreys, ravens, and owls; hawks, swans, pelicans, and storks; herons and bats and "all fowls that creep"; weasels, mice, tortoises, and ferrets; chameleons, lizards, snails, and moles. (Grasshoppers, for some reason, are just fine — but only in Yemen.) Jews who keep kosher still follow Leviticus to the letter, and anthropologists have made the book a favorite case study. Yet despite many elaborate explanations ranging from food safety to cultural xenophobia, no one has quite puzzled it out.

Except, perhaps, Calvin Schwabe. In 1979 Schwabe took a look at all this irrational smirking and gagging and decided it had gone far enough. As a veterinarian at the University of California, Davis, with years of experience in the Third World, he had seen too many people starving in places where food was everywhere to be found. Schwabe's answer was admirably practical: he wrote a cookbook. The most disgusting cookbook ever written. *Unmentionable Cuisine* gathered nearly four hundred recipes from all over the world — from baked bat and stuffed dormouse to stewed cat and Cajun muskrat. The only criterion was that the recipes offend someone. (Turkey testicles are apparently a fail-safe dish, as long as diners don't know what they're eating.)

Schwabe's aim was less sensationalist than revolutionary. "Some 3,500 puppies and kittens are born every hour in the United States," he wrote, "and the surplus among them represents at least 120 million pounds per year of potentially edible meat now being totally wasted." When thousands of Americans go hungry every year, and when the Romans considered suckling puppy a dish "fit for the gods," Schwabe's conclusion was obvious: our prejudices don't just define us; they can kill us.

Back in March, before I'd fully hatched my Strange Southern Foods tour, I put a call in to the Big Canoe, Georgia, satellite office of the Southern Foodways Alliance. Where could I find a chef, I asked, who specializes in "low-country cooking," the sophisticated cuisine woven together from African, American Indian, and European sources that plantation owners like Jefferson Davis once enjoyed? "Oh, dear," the woman at the other end said. "There aren't many places like that left." After a pause she directed me to the Horseradish Grill, in the heart of Buckhead, one of Atlanta's gilt-edged eastern neighborhoods. I called David Berry, the Grill's thirty-two-year-old chef, and asked him if he would make me a meal. There's one catch, I added. I'll be bringing the meat.

And so, on a cool, sunny Friday afternoon, completing the circle from Georgia to Alabama and back, I find myself waiting on the loading platform in back of the Grill, holding a bloody Ziploc bag full of bullfrog, wild hog, turtle, and possum. Berry emerges in chef's whites, his red beard closely trimmed, his manner crisp and professional. He grabs the bag, flips it over, squeezes it pensively, and examines the turtle claws, jutting under the plastic, with a herpetologist's eye. "All right," he says. "We'll do it up nice for you. Maybe add some southern vegetables." Then he smiles: "I hope you've got an open mind."

Five hours later, after a shave and a shower and a quick change into my least rumpled traveling clothes, I take my seat in Chef Berry's dining room. The decor could be called Nouvelle Hunting Lodge: the ceiling is vaulted, a fire blazes in the corner, mullioned windows are set off by bright oils on the wall, and jazz wafts down from hidden speakers. "Let me suggest a Sancerre," the sommelier says. "It's a great summer wine from the Loire Valley, with a bit of tang to stand up to what you'll be eating." Then he laughs, despite himself, and leans in closer. "The truth is I have no idea what to suggest. About the only thing I've done with frogs is flatten 'em with a post as a kid."

That thought is lodged in my mind when the first course arrives, but it's hard to connect it with the thing on the plate. On the one hand, no meat is quite so luridly anatomical as a pair of frog legs. Stripped of their slinky tights, every ligament and tendon, muscle and articulated joint looks ready to leap across the room. But the taste is worlds away from the swamp. Tender and buttery, with a subtle, amphibian chew, it's so mild the Sancerre almost overwhelms it. "I sautéed it for three or four minutes and then drizzled it with a lemon-caper, brown-butter sauce," Berry explains, settling in next to me. "Frog doesn't need a lot more than that."

Next up is the turtle soup, and with it a more jolting image: Ken Holyoak, the frog and turtle farmer, slouching against his truck at dusk a few days earlier, watching as I try to butcher a six-pound softshell. "You ever cleaned one of them before?" he asks me. "That's one nasty job." Half an hour and two broken knives later, I'm carving out the meaty thighs, prying open the joints, and reaching in to tear out the viscera. By the time I'm done, I'm actually humming to myself. The tune is "Passionate Kisses."

If the frog legs were a still life, this soup is pure abstraction. The broth is moss green and perfectly limpid, scattered with flakes of scallion and cubes of white and brown protein. There's no trace of ornery musculature, just a rich, tranquil flavor — a mixture of brine and fern and slumbering beast, as ancestral as chicken soup. "It's an old delicacy," Berry says, "so I didn't try to get too fancy. I just soaked the meat for a while, to pull out some of the blood, then simmered it in onions, garlic, celery, and a little carrot."

It's tasty, but Berry still wouldn't serve it in the restaurant. The problem is partly cultural and partly political. Like every other state, Georgia forbids restaurateurs to serve game or freshwater fish unless it's raised on a farm. "I can serve Chilean sea bass any day," Berry says. "But if I serve largemouth bass from a local lake, the health department could shut me down."

If Berry can't serve an acknowledged classic like turtle soup, what hope is there for my last dish?

The plate arrives looking like a hillbilly coat of arms: a proud possum shank emblazoned on a shield of grits, flanked by asparagus fleurs-de-lis and chevrons of wild hog tenderloin. "Gusta Plus Possium," the motto above it might read. Up until now I've tried to stick to things that a modern diner might reasonably adopt, under the right circumstances. But this possum has me worried. On the day that he killed it, Clark assured me that its fat was 100 percent polyunsaturated. "It'll clean your arteries like a Roto-Rooter," he said. But mostly I remember the possum's inscrutable, prehistoric face, its way of hissing and spitting when cornered, and its long, naked tail. "That's a natural air conditioner!" Clark explained. "The possum licks its tail, the blood circulates through it, and then the cool air cools it off." Somehow that image made it no more appetizing.

I start out slowly, with Berry watching my every facial tick. First the grilled hog — tender, smoky, and more flavorful than any pork loin I've ever tasted. Then the vegetables and grits (the latter a creamy revelation). And then, finally, the shank: big as my foot and dripping with thick possum gravy. I hold my breath at first but then slowly break into a smile. The gravy is strong and gamy, but with an uncomplicated charm, like something a cowboy might get served from a chuck wagon, and the meat is meltingly tender. "It's all in the preparation," Berry says. "I braised it very, very slowly in a little veal and chicken stock. I put in some onion, carrots, celery, mushrooms, and red wine, too. But that taste is all possum."

All possum. The words trigger an odd reaction in my mouth. It starts with the texture: fluttering pockets of fat are interleaved throughout the muscle fibers. Rubbery and slick, they bring to mind countless childhood dinners when my parents made me eat the fat from my pork chops. Then there's the aftertaste: that feral, faintly glandular presence rising through the sauce. This is an ancient animal, it tells me, one that was scurrying through primeval underbrush long before my ancestors, or their taste buds, had even evolved.

*

A solipsist might conclude that taste is all in the mind, but that's too easy. Our taste buds are just chemical receptors, designed to detect sweet and sour, salty and bitter, and no amount of prejudice can make them call a lemon sweet. Every taste is a story, a mystery for our minds to solve. Depending on the taster, the result may be tragedy or farce, hors d'oeuvre or abomination. Or, if we're lucky, something beyond categorization altogether.

A few years ago my father-in-law was driving to Nebraska to visit his ninety-year-old mother. He was fiddling with the radio dials, he says, looking down for just a second, when something hit the windshield with a terrific crash. Being a man of steady nerves and stoic Swedish character, he calmly maneuvered the car to the side of the highway and climbed out. There, lying in the road, was a wild turkey. He stared at the bird. The bird stared back. Then he picked it up and threw it in the trunk. No sense letting a thing like that go to waste.

When he arrived, he handed the bird over to his mother. She took it without comment — like him, she'd seen stranger things on the farm growing up. But that night, when he was dressing the bird, he found a surprise. Reaching inside like a magician, he pulled out an egg the size of his fist, still intact.

"What about that?" she asked him.

"Fry that up for breakfast," he said.

Long before I went to Georgia, that story served as a blunt reminder of my own culinary prejudices. I might think of myself as an adventurous eater, but my tastes had their limits, too: the bird I would eat; the egg, never.

Now I'm not so sure. Had I joined my father-in-law for breakfast that day and not known where the egg came from, I would have eaten it over easy or sunny side up. But he would have had the better meal. For an egg, eaten without prejudice, is like any other under the sun. But an egg with a story behind it, whether of a people, their history, or of a turkey crossing the road — that egg tastes like nothing you've ever imagined.

K. C. COLE

Mind Over Matter

FROM *The Los Angeles Times*

YOU JUST NEVER KNOW what the universe is going to be up to next. Why, just this month, a big glob of matter sitting out in space bent light like a giant lens, bringing into focus a far-off baby galaxy just coming into being. Not a month before that, the black hole at the center of our galaxy let out a loud belch of X rays, the best evidence yet that such a monster was feeding voraciously — proof from the belly of the beast, as it were.

This is only the latest in a long string of surprises; it sure makes you wonder what else the universe has in store. A paltry few hundred years ago, people believed there was nothing beyond what they could see with their eyes. You can imagine the astonishment that followed the invention of the first primitive microscope — the unnerving introduction to the teeming population of microbes that lives within (and on) our skin. "Who would have dreamed," writes Arthur C. Clarke, "that a tube connecting two lenses of glass would pierce the swarming depths of our being, force upon us incredible feats of sanitary engineering, master the plague, and create that giant upsurge out of unloosened human nature that we call the population explosion?"

Up until the start of this century, people thought that atoms were useful only as models because, of course, everyone knew that atoms could never be seen. Then Einstein showed that they not only *could be* seen, but *had been* seen, knocking about plant spores floating on water.

In short order, people learned not only to see atoms, but to look inside. Ernest Rutherford discovered the atomic nucleus when he bombarded gold atoms with particles streaming from radioactive rocks. Most of the particles passed right through, but some — surprisingly — were scattered backward. Rutherford wrote: "It was quite the most incredible event that ever happened to me in my life. It was almost as incredible as if you fired a fifteen-inch shell at a piece of tissue paper and it came back and hit you."

People had also assumed, with good reason, that it was impossible to know the composition of stars, since it was hardly possible to go and collect a sample. Then lines were found in starlight that told not only their makeup, but also their temperature, age, and motion — a kind of quantum mechanical bar code that reveals everything but their price. Long before Oprah, stars were spilling the intimate details of their lives to all who cared to listen. In fact, it was by decoding starlight that astronomers discovered (surprise of surprises) that 90 percent or more of the matter in the universe is unseen and perhaps unseeable.

Many discoveries have been so surprising that people didn't believe them, even when the evidence (like the jostling of plant spores) was right in front of their eyes. When Marie Curie discovered radioactivity, she speculated that the phenomenon might be evidence that atoms were actually disintegrating. This was inconceivable at a time when atoms were considered the indestructible building blocks of nature. Atoms were forever. Her idea seemed so bizarre that she decided not to publish. And what now seems an obvious fact — that the continents of Africa and South America fit together like puzzle pieces and were clearly once joined — was completely dismissed because no one believed that continents could move.

True, some discoveries are not so surprising. Recently, studies conducted with positron emission tomography (PET scans) revealed that the brains of teenagers are completely unlike those of adults. "Duh!" a parent (or teenager) might be tempted to say. But the very fact that positrons, which are a form of antimatter, could be used for medical purposes is pretty darned surprising — not to mention the fact that there is such a thing as antimatter at all.

We don't even know what it's possible to know. Today, some scientists say we can't know what lies beyond (or before) the Big Bang; others think they know how to look for evidence. String theory has been dismissed as so much pretty mathematics, since it can't be experimentally tested. Yet this month the University of Chicago is hosting a symposium on, yes, experimental tests of string theory.

In *The Unexpected Universe,* the naturalist Loren Eiseley tells of coming upon a spider in a forest, spinning the sticky spokes of the web that extend her senses out into the world. Just so, humans with their scientific senses have spun a web that reaches far beyond our ears and eyes. And like the spider, we lie "at the heart of it, listening." Yet Eiseley is even more impressed at what the spider *cannot* perceive. "Spider thoughts in a spider universe — sensitive to raindrop and moth flutter, nothing beyond. . . . What is it we are a part of that we do not see?"

Whatever it is, it's sure to be unexpected.

●

RICHARD CONNIFF AND
HARRY MARSHALL

In the Realm of Virtual Reality

FROM *Smithsonian*

IN THE COURTYARD of a monastery somewhere in central Bhutan, cockerels strut across the lattice of flagstones. A dozen temple dogs doze fitfully in the sun. A novice late for prayers scurries, sandals in hand, and vanishes into a doorway. Inside, it's cold and dark. The only light oozes down like buckwheat honey through a narrow atrium, dimly illuminating the weird figures painted on the heavy timbers, and the monstrous antlers of a supposedly extinct deer, the shou, lashed to a balcony railing.

From a black recess up ahead comes the low, sonorous sound of chanting, rising steadily in pitch and fervor. A horn made from a human femur goes *oh-woe-oh-oh-oh.* Cymbals stutter. A pair of short metallic horns put up a high-pitched, reedy sound. A drum beats. Our guide leads us into a narrow hall and then, after we remove our shoes, into the prayer room, where monks in maroon robes sit around the periphery and a smudge of incense drifts up from a censer. As our eyes adjust, strange animal shapes form vaguely out of the darkness: the dust-cloyed head of a tiger hanging on a wall, a huge, primitive-looking fish with long bony scales, a string of human hands. And off to one side of the altar, the thing we have come halfway round the world to see, a hanging figure with a white veil draped over its head.

Legend says a holy man brought Buddhism to the Himalayan kingdom of Bhutan twelve hundred years ago, flying in on the back of a tigress. Today, you fly into Paro International Airport in a Druk Air seventy-two-seat jet. It's reputed to be the most technically difficult landing in the world. "While flying in," the pilot announces on the initial descent, "you may find yourself coming closer to the terrain than is usual in a jetliner. Much closer. This is normal. Don't be alarmed." Most pilots crash several times before learning how to thread their way down the corridor of valleys to the airstrip. Fortunately, they make their mistakes on a flight simulator before attempting an actual approach. Bhutan has always been a place where the virtual and the real happily coexist.

Having opened itself to the outside world only forty years ago, the "Land of the Thunder Dragon" is also a place where tradition still shapes everyday life. In the terraced rice paddies of the Paro Valley, families thresh rice by hand, the sheaves swinging overhead, sending up plumes of dust, then down, *swot,* on a rock, over and over, until all the dry kernels of rice break loose and rain down in heaps. Chile peppers, another great national food, are spread out in bright red carpets on rooftops and hillsides to dry in the sun. Only about 700,000 people live in Bhutan, most of them in the fertile valleys. Uphill, beyond the last timbered farmhouse, the forest still covers 70 percent of a country the size of Switzerland (only more mountainous). Tigers, leopards, and bears still wander there. So, too, according to legend, does the migoi, which is what the Bhutanese call the yeti.

Migoi literally means "wild man," and the idea that there might be an undiscovered primate, a hairy quasi-human biped, still living in Bhutan's uncharted mountains is, according to some members of our expedition, the stuff of great adventure. And to some of us it is utter bunk. We agree, at least, that seeking the migoi is a way to get beneath the surface of this intriguing culture, which is why a British-American television partnership has sent us here. The group includes an Oxford-trained evolutionary biologist, a primatologist who has spent years working with monkeys in West Africa, and a British technical wiz who will keep our gear in working order. We have come equipped with camera traps, plaster of paris for casting footprints, and laboratory jars for sending back hair, scat, or tissue samples to be identified by DNA analysis. We've also got video cameras to record what the migoi means to the Bhutanese themselves. Our guide is Dasho Palden Dorji, a tall, chiseled thirty-seven-year-old who was educated at the University of California, Santa Barbara, and speaks English like an American. He's a true believer in the migoi, but patient with skeptics. Despite his easygoing manner, locals tend to trot when he issues an order. *Dasho* is a term of respect similar to "lord," and he is a first cousin to the king.

Other Bhutanese aren't so patient. When we note the failure of numerous previous attempts to find any hard evidence for the existence of the migoi, a forestry official wryly observes that Westerners only believe a species exists if a white man has given it a Latin name. In the 1950s a "new" primate was "discovered" in southern Bhutan and named with the help of a British naturalist, E. P. Gee. The golden langur, a gregarious monkey living in groups of fifteen or sixteen individuals, had been well known to the Bhutanese for centuries. But it is now immortalized in scientific literature as *Presbytis geei,* a name that still rankles some people in Bhutan. Local knowledge also got short shrift when yak herders reported seeing tigers in the mountains; science knew that tigers never go much above 6,000 feet. Then Bhutanese wildlife officials photographed a tiger crossing through a meadow at nearly 10,000 feet. So when locals who travel in the mountains say that a shy, solitary primate survives undetected there, isn't it a little arrogant to dismiss it as myth?

Maybe so, but our first few rounds of inquiry go to the skeptics. In a traditional farmhouse a few miles from the airport, a ritual purification ceremony is under way. The air is heavy with the smell of burning juniper and the sound of monks praying to drive unwanted spirits into intricate traps made of bamboo and ribbon. Dorji Tshering, the head of the household, sits cross-legged on the floor, fanning flies from his bowl of butter tea as he recalls his encounter with the migoi. The memory still gives him nightmares, though it happened fifty years ago. He and a friend had climbed the mountain behind the house in search of a suitable tree to saw into planks. Their journey took them through a glen called Migoi Shitexa and up to a shelter called Bandits' Cave. That night, as they collected firewood, they noticed strange footprints in the fresh snow. Tshering, now in his eighties, bow-legged, with a shock of white hair, taps his forearm at the elbow and then his knuckles, to show the length of the footprint. He knew what bear footprints looked like, he says, and this wasn't a bear. In the darkness, the two men heard the creaking and breaking of bamboo, followed by an eerie monotone call. Tshering imitates the sound. "This was definitely the migoi," he tells us. "We thought it was going to eat us up." "But what did you actually see?" we ask, and the answer is little more than a shape moving beyond the light of their fire, the size of a man, hairy, on two legs, its features indistinct in the darkness.

The stories told about the migoi are often like that, full of conviction but woefully short on detail. Or chockfull of plausible details, until a trapdoor suddenly drops open in the argument: a nomad digresses knowledgeably about the differences between migoi scat, which he says mainly contains bamboo, and bear scat, which is full of acorns. Then he adds that if you happen to go into the mountains when you are spiritually unclean, the migoi will bring typhoons and hailstorms down on your head. This odd mix of being so savvy about the natural world and yet so credulous is a little hard for outsiders like us to fathom. When we ask Tshering what "Migoi Shitexa" means, he says, "the place where the yeti scratches for lice." It is normal in Bhutan to be earthy and otherworldly at the same time.

The purification ceremony ends late in the day when the children carry the spirit traps out of the house and into the fields. The traps are laden with bread, fruit, money, and strands of fabric to placate the evil spirits now bottled up within. "Ghostbusters for Buddhists," someone remarks. But in truth, the whole scene feels as if we've been set down in the middle of a Brueghel painting, in the Europe of five hundred years ago: blue smoke in the air, a neat golden stack of rice straw in the farmyard with a straw finial on top, a pervasive sense of religious faith, and a kind of ribald peasant contentment with the course of life. Standing outside, we can hear the monks raucously chanting their last few prayers. One of them turns his weathered face over his shoulder, grinning and shouting the words of the prayer out to us through a tiny arched window. A few minutes later, a mangy dog comes trotting back to the house from the field, a strand of ribbon trailing over its ears from its raid on the contents of a spirit trap.

The winding, one-lane mountain roads take us into central Bhutan, where the monastery called Gangtey Gompa is famous for two things: black-necked cranes forage in the broad wetlands below, and within the walls of the monastery itself is a mummified *mechume,* the putative remains of a small yeti.

The black-necked cranes are real enough, and by late October their *trum-trum* calls echo across the frosted marshes. In the brilliant white light of dawn, groups of them soar down, the sun glinting off their six-foot wingspans. In a country where the human life expectancy is about fifty years, the cranes live to be nearly half that and are revered as bodhisattva, or Buddhist deities. It's said that at the time of their departure each year, before their migration north to the Tibetan plateau, the cranes fly three times around the monastery, the clockwise ritual of any Buddhist pilgrim. There's a simpler explanation: ornithologists say the cranes are simply flying around trying to gain altitude. But for us, the symbolic and supernatural values are starting to become more intriguing. Coming from a world where we see the landscape in terms of lots and subdivisions, we're learning to envy the way people in Bhutan still tell stories about their own hills and valleys.

It's easy to get caught up in local values, especially within the darkened monastery itself, where the monks chant *Om ma ni pad me hung* ("Praise to the jewel in the lotus"), as they have chanted numberless times since the twelfth century. Our guide, Dasho Palden, says no Westerner has ever entered the monastery's inner sanctum before, but the abbot has arranged this visit especially to show us the mechume. The atmosphere in the prayer room is somewhere between a reliquary and a rag-and-bone shop. By the flickering butter candles, we step gingerly around sacks hanging like punching bags from the rafters. Asked what's in them, Palden replies, "Diseases. If they get out, they will spread." The wide wooden floorboards have been polished by generations of barefoot pilgrims prostrating themselves before the crowded altar. Dasho Palden prostrates himself, too, and then, according to custom, a monk hands him three dice to foresee the fate of the expedition we are about to undertake into the perilous high country. Palden rolls a thirteen. In the Eastern scheme of things, he assures us, this is a good, solid number.

Afterward, he leads us over to the far side of the altar, to examine the mechume. The story goes that roughly two hundred years ago, in a village two days' hike from here, a series of killings occurred. A local holy man determined that this mechume was the culprit. He tracked it down and cut it in half with his sword, and the corpse has been hanging at Gangtey Gompa ever since. The head, no larger than a child's, slumps down, chin resting on the sack of skin that was once its chest. The withered hands and feet hang by threads and bits of dried flesh. The eyes are squeezed shut, and the mouth is stretched wide. Dasho Palden suggests that the mouth is making the sound of the mechume, *Woooooo.* But to us it looks as if the mouth has been frozen at the split second of this creature's death in an eternal wail of agony and despair. Either way, someone has stuffed an offering of *ngultrum,* the local currency, into its mouth.

That face continues to haunt us for days afterward, and one night around the campfire, the skeptics in our group develop an alternative theory about how it came to be at Gangtey Gompa: imagine a rural holy man beset by angry, terrified villagers demanding action. Was it practical to think he could have caught and killed a creature as elusive as the mechume, which the Bhutanese themselves sometimes describe as part god, part devil? Or did he simply find some poor scapegoat, a loner, a madman, someone who would not be missed — much as New Englanders once burned ordinary women as witches? In any case, we agree that what is hanging at Gangtey Gompa is unmistakably, unbearably, human.

"You'll have to excuse me," Kunzang Choden says, when she greets us at the front door of her family compound, "but I am expecting my uncle's reincarnation. He's an American boy, fourteen now." He's due to arrive sometime before nightfall, and everyone's bustling around getting ready. Ugyen Choling Palace, where Choden's ancestors have lived since the fifteenth century, is at the end of the long, idyllic Tang Valley in central Bhutan. It's two hours by car from the nearest town, Bumthang, and a one-hour hike uphill from the nearest road. The palace once lorded over the valley, but almost all the rooms are empty now, with the wooden window panels drawn shut to keep out pigeons. Choden, who is forty-eight, grew up before electricity, television, and videotapes came to Bhutan, and the greatest pleasure of her childhood was listening to yak herders, back from the wild, telling their stories of the high country. She's now Bhutan's leading folklorist, and she's put together an anthology called *Bhutanese Tales of the Yeti.*

Choden accepts our impression that the migoi stories are much like European fairy tales. The migoi, like an ogre in a fairy tale, often comes to a gruesome end. "I also have my sympathies with the migoi," Choden admits. "I'm sure if the migoi could tell its story, these terrible things wouldn't happen." But humans tell the stories, and "even though we Bhutanese live in nature, we have to be able to be the masters somehow."

Is it possible that's why Bhutan still has its migoi stories? Because there's still wilderness, unmastered, just beyond the farmyard gate, much as there was dark forest in Europe back when the fairy tale tradition there was strong? Choden replies that urbanized Western countries still have their credulous tales, only updated a little: "*Star Wars* and all that comes from the fact that they've lost the wilderness and they have to look beyond, whereas we still have our environment intact, and we know that there are spirits living under the trees, spirits residing on the mountaintops. So this is still a part of our reality."

As we talk, her pet Lhasa apso noses around. "Ignore him," she advises. "He is a most terrible attention-seeker." And then she continues, "The supernatural and the natural, we do not delineate. People in the villages still perform rituals to appease the spirits that they have harmed, knowingly or unknowingly. We have, I guess, a ritual world. We have so much that is not explained. And we do not want it explained. What I fear most is that soon, with our children all going into Western schools and learning more about Western culture and beliefs, this will be lost."

Then, since our expedition is about to head off into the mountains, seeking explanations and evidence, she offers some parting advice: if we meet a female yeti, keep in mind that the yeti's long, sagging breasts make her top-heavy on a down slope. "So, yes, run downhill," Choden says. Then she grins and adds, "Go down the cliff, I think."

Next night, when our packhorses have been set free to graze, and our tents are staked down, the group gathers for dinner. We're in bear and tiger country now, our evolutionary biologist reminds us, and he gives each of us an "attack alarm." "It will produce a 107-decibel shriek, and that should scare off just about any animal." Then he pushes the trigger.

"The horses didn't twitch," someone says, when we have recovered enough to hear them contentedly tearing up the grass just beyond the light of our lanterns. We decide to put our faith instead in one of our guides, known simply as "303," for the battered old .303 Enfield rifle he keeps slung over one shoulder. The government has provided us with this armed protection because unpredictable Himalayan black bears often maul or kill unlucky yak herders.

We camp in an open meadow, but there's still forest all around. On the big old spruce trees, Spanish moss drifts in the breeze like an old man's wispy beard. Rhododendrons with leaves as big as fans grow in the understory, along with stunted little bamboo. Soft green moss carpets the floor and swells up in pom-pom clumps over broken branches. The forest itself feels as though animals — and spirits, too — have somehow taken root in the steep mountain soil. But our camera traps, with motion sensors and infrared gear for shooting by night as well as by day, turn up no big, dangerous animals. The movement of the vegetation sets them off. And once, some creature leaves its tantalizing shadow across a corner of the video frame.

Proving a species exists is relatively straightforward. The rule of habeas corpus applies: you must present a body or specimen, a holotype that other scientists can examine in a laboratory setting. Proving that a species doesn't exist, on the other hand, is near to impossible. This dawns on us, literally, at Rodong La, at 14,000 feet on the trek from Bumthang to Lhuentse.

According to Kunzang Choden, this lonely pass is one of the most frequent sites for yeti spottings. So one night two of our party, the skeptics, camp out alone at Rodong La, without 303 or any other protection, offering ourselves to the yeti as unbelievers. We spend the first part of the night holed up in a blind we've pitched in a shadowy stand of gnarled rhododendrons. It's cold and lonely. A cloud envelops us, and we have no moon or stars to give us the minimal light needed for the image intensifier on our video camera. Everything is damp and eerily silent, with no birds or insects. It feels as if the cloud has somehow sucked the sound out of the sky. On this remote spot, one of Choden's informants once saw a yeti face to face with a tiger.

We have no such luck. But when the lichen and moss have gone brittle with frost, the cloud lifts. The faint yellow light before dawn unveils the landscape around us, and then other hills and mountains, which gradually separate from one another as the sun rises. In the distance, brilliant orange light catches on the snowcapped peak of Gangkar Punsum. At 25,000 feet, it is the highest mountain in Bhutan, and no one has ever climbed it.

By now, we are out of the blind, wandering in amazement. To the northeast, a wide lazy river of clouds flows through a valley, and at the end, the clouds spill down over a precipice like a waterfall, tumbling and steaming between two spruce promontories. Beyond that, other valleys, still forested and without people, delve toward infinity. It makes us realize that in this vast unexplored landscape anything is possible.

But not necessarily right now. In all our climbing up lonely mountain passes and down endless winding stairways built into the cliff face, we turn up nothing remotely like a yeti. The camera traps yield one squirrel, and a yak herder urinating by the side of the trail. Then one day we make the steep, sweaty descent into a valley, where the white noise of a river rises up to meet us, along with the cries of farm children. In the remote, roadless village of Khaine Lhakhang, we meet a man named Sonam Dhendup. He's thirty-seven years old, with black hair just starting to go gray, a wispy goatee, and short, rough, muscular legs, the product of a lifetime in the Himalayas.

Dhendup tells us he has worked for the past twelve years as a migoi-spotter for the government. He has yet to see a migoi himself. But he knows a place two days' hike away, where the migoi comes to eat bamboo each spring and where his droppings pile up in heaps. The season is wrong. But the mountains are tempting, and Dhendup tells us enough to convince us he knows the local wildlife.

Accompanied by 303, two of our party follow Dhendup back up until they are walking among the clouds at 12,000 feet. In the dense, sodden forest there, Dhendup kneels to indicate the fresh pugmarks of a large male tiger, a few minutes ahead. The worn bolt on the .303 Enfield slides home with a sharp click. A soft rain begins to fall. As Dhendup creeps barefoot through the forest, he points out more tiger prints, some wild boar grubbings, places where bears have clearly rubbed against trees. He's a connoisseur of Himalayan wildlife, able to distinguish black bear and brown bear by their paw marks alone. Why would he, as skeptics like to suggest, mistake either species for a yeti?

The hill fog closes in around them, and the bamboo forest becomes a dripping prison of crisscrossed stems and pungent leaf mold. They circle aimlessly for hours, until Dhendup veers off purposefully and they arrive at the hollow tree he has been seeking. As expected, the dung heaps he saw there last spring are long gone. But there's a cavity in the tree just large enough for a human, or some creature of similar size, to hunker down and find refuge. A careful inspection turns up hairs stuck in the rough sides of the cavity. Some of the strands are long and dark, others shorter and more bristly. Not bear, says Dhendup. The hairs are, at the very least, worth packing up and taking home for DNA analysis.

Months later, in a laboratory at Oxford University, geneticist Bryan Sykes is pleased with the samples we have brought back. The hairs have plump follicles, the part that contains the DNA. Prospects for identifying the source of these hairs seem good. Sykes starts with the plausible assumption that he is looking at bear DNA, or maybe wild pig. But the DNA seems to suggest otherwise. His laboratory simply cannot sequence it. "We normally wouldn't have any difficulty at all," says Sykes. "It had all the hallmarks of good material. It's not a human, it's not a bear, nor anything else that we've so far been able to identify. We've never encountered any DNA that we couldn't sequence before. But then, we weren't looking for the yeti." It is, he says, "a mystery, and I didn't think this would end in a mystery."

What a scientist cannot add is that sometimes a mystery is enough. The skeptics in our group had conceded as much that morning back in Rodong La. Watching the sunrise on the endless mountains, we agreed that if we had a choice — a land rich in wilderness, and rich in demons, too, or a land "civilized" and full of skeptics like us — we would gladly take Bhutan the way it is, yeti and all.

FREDERICK C. CREWS

Saving Us from Darwin

FROM *The New York Review of Books*

IT IS NO SECRET that science and religion, once allied in homage to divinely crafted harmonies, have long been growing apart. As the scientific worldview has become more authoritative and self-sufficient, it has loosed a cascade of appalling fears: that the human soul, insofar as it can be said to exist, may be a mortal and broadly comprehensible product of material forces; that the immanent, caring God of the Western monotheisms may never have been more than a fiction devised by members of a species that self-indulgently denies its continuity with the rest of nature; and that our universe may lack any discernible purpose, moral character, or special relation to ourselves. But as those intimations have spread, the retrenchment known as creationism has also gained in strength and has widened its appeal, acquiring recruits and sympathizers among intellectual sophisticates, hard-headed pragmatists, and even some scientists. And so formidable a political influence is this wave of resistance that some Darwinian thinkers who stand quite apart from it nevertheless feel obliged to placate it with tactful sophistries, lest the cause of evolutionism itself be swept away.

As everyone knows, it was the publication of *The Origin of Species* in 1859 that set off the counterrevolution that eventually congealed into creationism. It isn't immediately obvious, however, why Darwin and not, say, Copernicus, Galileo, or Newton should have been judged the most menacing of would-be deicides. After all, the subsiding of faith might have been foreseeable as soon as the newly remapped sky left no plausible site for heaven. But people are good at living with contradictions, just so long as their self-importance isn't directly insulted. That shock was delivered when Darwin dropped his hint that, as the natural selection of every other species gradually proves its cogency, "much light will be thrown on the origin of man and his history."

By rendering force and motion deducible from laws of physics without reference to the exercise of will, leading scientists of the Renaissance and Enlightenment started to force the activist lord of the universe into early retirement. They did so, however, with reverence for his initial wisdom and benevolence as an engineer. Not so Darwin, who saw at close range the cruelty, the flawed designs, and the prodigal wastefulness of life, capped for him by the death of his daughter Annie. He decided that he would rather forsake his Christian faith than lay all that carnage at God's door. That is why he could apply Charles Lyell's geological uniformitarianism more consistently than did Lyell himself, who still wanted to reserve some scope for intervention from above. And it is also why he was quick to extrapolate fruitfully from Malthus's theory of human population dynamics, for he was already determined to regard all species as subject to the same implacable laws. Indeed, one of his criteria for a sound hypothesis was that it must leave no room for the supernatural. As he wrote to Lyell in 1859, "I would give absolutely nothing for the theory of Natural Selection, if it requires miraculous additions at any one stage of descent."

Darwin's contemporaries saw at once what a heavy blow he was striking against piety. His theory entailed the inference that we are here today not because God reciprocates our love, forgives our sins, and attends to our entreaties but because each of our oceanic and terrestrial foremothers was lucky enough to elude its predators long enough to reproduce. The undignified emergence of humanity from primordial ooze and from a line of apes could hardly be reconciled with the unique creation of man, a fall from grace, and redemption by a person of the godhead dispatched to earth for that end. If Darwin was right, revealed truth of every kind must be unsanctioned. "With me the horrid doubt always arises," he confessed in a letter, "whether the convictions of man's mind, which has been developed from the mind of the lower animals, are of any value or at all trustworthy. Would any one trust in the convictions of a monkey's mind . . . ?"

In a sentence that is often misconstrued and treated as a scandal, Richard Dawkins has asserted that "Darwin made it possible to be an intellectually fulfilled atheist." What he meant was not that Darwinism requires us to disbelieve in God. Rather, if we are already inclined to apprehend the universe in strictly physical terms, the explanatory power of natural selection removes the last obstacle to our doing so. That obstacle was the seemingly irrefutable "argument from design" most famously embodied in William Paley's *Natural Theology* of 1802. By showing in principle that order could arise without an artificer who is more complex than his artifacts, Darwin robbed Paley's argument of its scientific inevitability.

With the subsequent and continually swelling flood of evidence favoring Darwin's paradigm, evolutionism has acquired implications that Darwin himself anticipated but was reluctant to champion. Daniel C. Dennett has trenchantly shown that the Darwinian outlook is potentially a "universal acid" penetrating "all the way down" to the origin of life on earth and "all the way up" to a satisfyingly materialistic reduction of mind and soul. True enough, natural selection can't tell us how certain organic molecules first affixed themselves to templates for self-duplication and performed their momentous feat. But the theory's success at every later stage has tipped the explanatory balance toward *some* naturalistic account of life's beginning. So, too, competitive pressures now form a more plausible framework than divine action for guessing how the human brain could have acquired consciousness and facilitated cultural productions, not excepting religion itself. It is this march toward successfully explaining the higher by the lower that renders Darwinian science a threat to theological dogma of all but the blandest kind.

That threat has been felt most keenly by Christian fundamentalists, whose insistence on biblical literalism guarantees them a head-on collision with science. They are the faction responsible for creationism as most people understand the term: the movement to exclude evolution from the public school curriculum and to put "creation science" in its place. The goal of such "young-earthers" is to convince students that the Bible has been proven exactly right: our planet and its surrounding universe are just six thousand years old, every species was fashioned by God in six literal days, and a worldwide flood later drowned all creatures (even the swimmers) except one mating pair of each kind.

Creation science enjoyed some political success in the 1980s and 1990s, packing a number of school boards and state legislatures with loyalists who then passed anti-Darwinian measures. Clearly, though, the movement is headed nowhere. Its problem isn't the absurdity of its claims but rather their patently question-begging character. "Findings" that derive from Scripture can never pass muster as genuine science, and once their sectarian intent is exposed, they inevitably run up against the constitutional ban on established religion.

But the ludicrous spectacle of young-earth creation science masks the actual strength of creationism in less doctrinaire guises. According to a recent poll, only 44 percent of our fellow citizens agree with the proposition "Human beings, as we know them today, developed from earlier species of animals." One of the dissenters may be our current president, who went on record, during the Kansas State Board of Education controversy of August 1999, as favoring a curricular balance between Darwinian and creationist ideas. His administration, moreover, is partial to charter schools, public funding of private academies, and a maximum degree of autonomy for local boards. If creationism were to shed its Dogpatch image and take a subtler tack, laying its emphasis not on the deity's purposes and blueprints but simply on the unlikelihood that natural selection alone could have generated life in its present ingenious variety, it could multiply its influence many fold.

Precisely such a makeover has been in the works since 1990 or so. The new catchword is "intelligent design" (ID), whose chief propagators are Phillip E. Johnson, Michael J. Behe, Michael Denton, William A. Dembski, Jonathan Wells, Nancy Pearcey, and Stephen C. Meyer. Armed with Ph.D.'s in assorted fields, attuned to every quarrel within the Darwinian establishment, and pooling their efforts through the coordination of a well-funded organization, Seattle's Discovery Institute, these are shrewd and media-savvy people. They are very busy turning out popular books, holding press conferences and briefings, working the Internet, wooing legislators, lecturing on secular as well as religious campuses, and even, in one instance, securing an on-campus institute all to themselves.

The IDers intend to outflank Darwin by accepting his vision in key respects, thereby lending weight to their one key reservation. Yes, most of them concede, our planet has been in orbit for billions of years. No, earth's 10 million species probably weren't crammed into Eden together. And yes, the extinction of some 99 percent of those species through eons preceding our own tardy appearance is an undeniable fact. Even the development, through natural selection, of adaptive variation within a given species is a sacrificed pawn. The new creationists draw the line only at the descent of whole species from one another. If those major transitions can be made to look implausible as natural outcomes, they can be credited to the Judeo-Christian God, making it a little more thinkable that he could also, if he chose, fulfill prophecies, answer prayers, and raise the dead.

This is, on its face, a highly precarious strategy. According to the premises that intelligent design freely allows, speciation *isn't* very hard to explain. If natural selection can produce variations without miraculous help, there is every reason to suppose that it can yield more fundamental types as well. Indeed, Darwin believed, and many contemporary biologists agree, that the very distinction between variation and speciation is vacuous. One species can be distinguished from its closest kin only retrospectively, when it is found that the two can no longer interbreed. The cause of that splitting can be something as mundane as a geographical barrier erected between two groupings of the same population, whose reproductive systems or routines then develop slight but fateful differences. And if one of those sets then goes extinct without leaving traces that come to the notice of paleontologists, the surviving set may not be considered a new species after all, since no discontinuity in breeding will have come to light. The whole business requires a bookkeeper, perhaps, but surely not a God.

In effect, then, the intelligent design team has handed argumentative victory to its opponents before the debate has even begun. As the movement's acknowledged leader, the emeritus UC-Berkeley law professor Phillip Johnson, concedes in his latest book, *The Wedge of Truth,* "If nature is all there is, and matter had to do its own creating, then there is every reason to believe that the Darwinian model is the best model we will ever have of how the job might have been done." Such a weak hand prompts Johnson and others to retreat to the Bible for "proof" that nature is subordinate to God. If scientists can't perceive this all-important truth, it's because their "methodological naturalism" partakes of a more sweeping "metaphysical naturalism" — that is, a built-in atheism. Once this blindness to spiritual factors becomes generally recognized, the persuasiveness of Darwinism will supposedly vanish.

While awaiting this unlikely outcome, however, ID theorists also make an appeal to consensual empiricism. The rhetorically adept Johnson, for example, highlights every disagreement within the evolutionary camp so that Darwinism as a whole will appear to be moribund. There are many such areas of dispute, having to do with morphological versus genetic trees of relationship; with convergent evolution versus common descent; with individual versus group selection; with "punctuated equilibria" versus relatively steady change; with sociobiological versus cultural explanations of modern human traits; and with the weight that should be assigned to natural selection vis-à-vis sexual selection, symbiosis, genetic drift, gene flow between populations, pleiotropy (multiple effects from single genes), structural constraints on development, and principles of self-organizing order. But Johnson misportrays healthy debate as irreparable damage to the evolutionary model — to which, as he knows, all of the contending factionalists comfortably subscribe.

The Wedge of Truth adds nothing of substance to Johnson's four previous volumes in the same vein. By now, though, his cause has been taken up by younger theorists whose training in science affords them a chance to make the same case with a more imposing technical air. In *Icons of Evolution: Science or Myth?*, for example, Jonathan Wells mines the standard evolutionary textbooks for exaggerated claims and misleading examples, which he counts as marks against evolution itself. His goal, of course, is not to improve the next editions of those books but to get them replaced by ID counterparts. More broadly, he calls for a taxpayer revolt against research funding for "dogmatic Darwinists" and for the universities that house their "massive indoctrination campaign." What he cannily refrains from saying is that a prior religious commitment, not a concern for scientific accuracy, governs his critique. One must open the links on Wells's Web site to learn that, after consulting

God in his prayers and attending to the direct personal urging of the Reverend Sun Myung Moon, whom he calls "the second coming of Christ," he decided that he should "devote [his] life to destroying Darwinism."

What is truly distinctive about the intelligent design movement is its professional-looking attack on evolution at the molecular level. Darwin had famously dared his critics to find "any complex organ . . . which could not possibly have been formed by numerous, successive, slight modifications." Having failed to unearth any such organ, anti-evolutionists have recently turned to the self-replicating cell, with its myriad types of proteins and its many interdependent functions. In *Darwin's Black Box: The Biochemical Challenge to Evolution* (1996), the Catholic biochemist Michael J. Behe has asked whether such amazing machinery could have come into existence by means of "slight modifications." His answer is no: God's intervention within the cell can be demonstrated through the elimination of every possibility other than conscious design. Without waiting to learn what his fellow biochemists think of this breakthrough (they have scoffed at it), Behe generously ascribed it to them and called it "one of the greatest achievements in the history of science."

The heart of Behe's case is his notion of irreducible complexity. Any mechanical or biological system — a mousetrap, say, or a bacterial flagellum — is irreducibly complex if each of its elements is indispensable to its functioning. How could one irreducibly complex system ever evolve into another? According to Behe, any stepwise mutation that altered the original would have rendered it not just clumsy but useless and thus incapable of survival. To maintain otherwise, he urges, would be like saying that a bicycle could grow into a motorcycle by having its parts traded, one by one, for a heavy chassis, a gearbox, spark plugs, and so on, while never ceasing to constitute a maximally efficient vehicle. Since that is impossible, Behe declares, "the assertion of Darwinian molecular evolution is merely bluster."

The IDers have closed ranks behind Behe as their David to the Darwinian Goliath. His inspiration pervades their manifesto anthology, *Mere Creation: Science, Faith and Intelligent Design,* a triumphalist volume in which the impending collapse of evolutionism is treated as a settled matter. In the view of the editor, William Dembski, Darwinism is already so far gone, and the prospect of reverse-engineering God's works to learn his tricks is so appealing, that "in the next five years intelligent design will be sufficiently developed to deserve funding from the National Science Foundation."

Dembski himself is the author of two books, *The Design Inference: Eliminating Chance Through Small Probabilities* (1998) and *Intelligent Design: The Bridge Between Science and Theology* (1999), that put the case for irreducible complexity on more general grounds than Behe's. The key question about Darwinism, Dembski has perceived, is the one that Paley would have asked: whether natural selection can result in organs and organisms whose high degree of order associates them with made objects (a compass, say) rather than with found objects such as a rock. By applying an algorithmic "explanatory filter," Dembski believes, we can make this discrimination with great reliability. Design must be inferred wherever we find *contingency* (the object can't be fully explained as an outcome of automatic processes), *complexity* (it can't have been produced by chance alone), and *specification* (it shows a pattern that we commonly associate with intelligence). Since living forms display all three of these properties, says Dembski, they must have been intelligently designed.

Working evolutionists, once they notice that Behe's and Dembski's "findings" haven't been underwritten by a single peer-reviewed paper, are disinclined to waste their time refuting them. Until recently, even those writers who do conscientiously alert the broad public to the fallacies of creationism have allowed intelligent design to go unchallenged. But that deficit has now been handsomely repaired by two critiques: Robert T. Pennock's comprehensive and consistently rational *Tower of Babel*, the best book opposing creationism in all of its guises, and Kenneth R. Miller's *Finding Darwin's God*, whose brilliant first half reveals in bracing detail that intelligent design is out of touch with recent research.

As Pennock shows, Behe's analogical rhetoric is gravely misleading. He makes it seem that *one* exemplar of a molecular structure faces impossible odds against transforming itself into *one* quite different form while remaining highly adaptive. But evolutionary change, especially at the level of molecules and cells, occurs in vast populations, all but a few of whose members can be sacrificed to newly hostile conditions and dead-end mutations. Antibiotic resistance among bacteria and the rapid evolution of the HIV virus are two common examples that carry more weight than any number of mousetraps and bicycles.

Both Pennock and Miller demonstrate that evolution is not a designer but a scavenger that makes do with jury-rigged solutions and then improves them as opportunities and emergencies present themselves. Typically, the new mechanism will have discarded "scaffolding" elements that were no longer needed. And conversely, a part that may have been only mildly beneficial in one machine can become essential to its successor, which may serve a quite different end. This chain of makeshift solutions is no less true of cilia and flagella than it is of the reptilian jaw that eventually lent two bones to the mammalian middle ear.

As for Dembski, his explanatory filter assumes what it is supposed to prove, that natural causes can't have brought about the "complex specified information" characteristic of life forms. Dembski fails to grasp that Darwinism posits neither chance nor necessity as an absolute explainer of those forms. Rather, it envisions a continual, novelty-generating disequilibrium between the two, with aleatory processes (mutation, sexual recombination, migratory mixing) and the elimination of the unfit operating in staggered tandem over time. Declaring this to be impossible by reference to information theory, as Dembski does with mathematical sleight-of-hand, is just a way of foreclosing the solid evidence in its favor.

By denying that natural selection can generate specified complexity, theorists like Dembski and Behe saddle themselves with the task of determining when the divine designer infused that complexity into his creatures. Did he do it (as Behe believes) all at once at the outset, programming the very first cells with the entire repertoire of genes needed for every successor species? Or did he (Dembski's preference) opt for "discrete insertions over time," molding here a Velociraptor, there a violet, and elsewhere a hominid according to his inscrutable will? Miller and Pennock show that both models entail a host of intractable problems.

*

The proper way to assess any theory is to weigh its explanatory advantages against those of every extant rival. Neo-Darwinian natural selection is endlessly fruitful, enjoying corroboration from an imposing array of disciplines, including paleontology, genetics, systematics, embryology, anatomy, biogeography, biochemistry, cell biology, molecular biology, physical anthropology, and ethology. By contrast, intelligent design lacks any naturalistic causal hypotheses and thus enjoys no consilience with any branch of science. Its one unvarying conclusion — "God must have made this thing" — would preempt further investigation and place biological science in the thrall of theology.

Even the theology, moreover, would be hobbled by contradictions. Intelligent design awkwardly embraces two clashing deities — one a glutton for praise and a dispenser of wrath, absolution, and grace, the other a curiously inept cobbler of species that need to be periodically revised and that keep getting snuffed out by the very conditions he provided for them. Why, we must wonder, would the shaper of the universe have frittered away 13 billion years, turning out quadrillions of useless stars, before getting around to the one thing he really cared about, seeing to it that a minuscule minority of earthling vertebrates are washed clean of sin and guaranteed an eternal place in his company? And should the God of love and mercy be given credit for the anopheles mosquito, the schistosomiasis parasite, anthrax, smallpox, bubonic plague . . . ? By purporting to detect the divine signature on every molecule while nevertheless conceding that natural selection does account for variations, the champions of intelligent design have made a conceptual mess that leaves the ancient dilemmas of theodicy harder than ever to resolve.

A conceptual mess can persist indefinitely, however, if its very muddle allows cherished illusions to be retained. As we will see, intelligent design is thriving not just among programmatic creationists but also in cultural circles where illogic and self-indulgence are usually condemned. And even stronger evidence that the Darwinian revolution remains incomplete can be found within the evolutionary establishment itself, where Darwin's vision is often prettified to make it safe for doctrines that he himself was sadly compelled to leave behind.

*

Above, I argued that "intelligent design" — the theory that cells, organs, and organisms betray unmistakable signs of having been fashioned by a divine hand — bears only a parodic relationship to a research-based scientific movement. In a world where empirical issues were settled on strictly empirical grounds, ID would be a doctrine without a future. But scientific considerations can take a back seat when existential angst, moral passions, and protectiveness toward sacred tradition come into play.

One doesn't have to read much creationist literature, for example, before realizing that anti-Darwinian fervor has as much to do with moral anxiety as with articles of revealed truth. Creationists are sure that the social order will dissolve unless our children are taught that the human race was planted here by God with instructions for proper conduct. Crime, licentiousness, blasphemy, unchecked greed, narcotic stupefaction, abortion, the weakening of family bonds — all are blamed on Darwin, whose supposed message is that we are animals to whom everything is permitted. This is the "fatal glass of beer" approach to explaining decadence. Take one biology course that leaves Darwin unchallenged, it seems, and you're on your way to nihilism, Eminem, and drive-by shootings.

Crude though it is, such an outlook is not altogether dissimilar to that of prominent American neoconservatives who see their nation as consisting of two cultures, one of which is still guided by religious precepts while the other has abandoned itself to the indulgences of "the sixties." Whatever the descriptive merits of that scheme, it exhibits the same foreshortened and moralized idea of causality that we see among the creationists. If the social fabric appears to be fraying, it's less because objective conditions have changed than because the very principles of authority and order have been gradually undermined by atheistical thinkers from Marx, Nietzsche, and Freud through Herbert Marcuse, Norman Mailer, and Timothy Leary. And Darwin, despite his personal commitment to duty, sometimes makes his way onto the enemies list as well.

The most articulate proponent of the "two cultures" theory is the distinguished historian Gertrude Himmelfarb, who also happens to be the author of a learned study of Darwin and his milieu, published in 1959. Like her husband, Irving Kristol, who has declared "the very concept of evolution questionable," Himmelfarb showed no patience with natural selection in her book. She aimed to prove that Darwin's "failures of logic and crudities of imagination emphasized the inherent faults of his theory. . . . The theory itself was defective, and no amount of tampering with it could have helped." Himmelfarb's *Darwin* remains an indispensable contribution to Victorian intellectual history, but its animus against Darwin and Darwinism makes the book read like a portent of the neoconservatives' realization that, by liberal default, they must be the party of the creator God.

In recent decades both Kristol and Himmelfarb have been ideological bellwethers for the monthly *Commentary,* which, interestingly enough, has itself entered combat in the Darwin wars. In 1996 the magazine caused a ripple of alarm in scientific circles by publishing David Berlinski's essay "The Deniable Darwin," a florid and flippant attack that rehearsed some of the time-worn creationist canards (natural selection is just a tautology, it contravenes the second law of thermodynamics, and so forth) while adding the latest arguments from intelligent design. And as if to show how unimpressed they were by the corrections that poured in from evolutionists, the editors brought Berlinski onstage for an encore in 1998, this time declaring that he hadn't been taken in by party-line apologetics for the Big Bang, either.

In answering his dumbfounded critics, Berlinski — now a fellow of the Discovery Institute in Seattle, an organization founded to promote anti-Darwinian ideas — denied that he is a creationist. What he surely meant, however, was that he isn't a *young-earth* creationist. His Darwin essay called Paley's 1802 argument from design "entirely compelling," leaving us with no reason to look beyond the following explanation of life: "God said: 'Let the waters swarm with swarms of living creatures, and let fowl fly above the earth in the open firmament of heaven.'" By his fellow anti-Darwinian Phillip Johnson's definition — "A creationist is simply a person who believes that God creates" — Berlinski is no less a creationist than every other member of the ID movement.

Commentary is not the only rightward-leaning magazine to have put out a welcome mat for intelligent design. For some time now, Richard John Neuhaus, editor of the conservative religious journal *First Things,* has been using Phillip Johnson as his authority on the failings of natural selection — this despite the fact that Johnson's willful incomprehension of the topic has been repeatedly documented by reviewers. On the dust jacket of *The Wedge of Truth,* furthermore, Neuhaus calls Johnson's case against Darwin "comprehensive and compellingly persuasive," adding, remarkably, that its equal may not be found "in all the vast literature on Darwinism, evolution, creation, and theism."

Further: when, in 1995, the neoconservative *New Criterion* sought an appropriate reviewer for Daniel C. Dennett's *Darwin's Dangerous Idea* — a book that rivals Richard Dawkins's *The Blind Watchmaker* as creationism's bête noire — it was Johnson again who was chosen to administer the all-too-predictable put-down. *The New Criterion*'s poor opinion of evolutionism can be traced to its managing editor Roger Kimball's esteem for the late philosopher David Stove, whose book *Darwinian Fairytales* (1995) is notable for its obtusely impressionistic way of evaluating scientific hypotheses. But since Kimball and *The New Criterion* regularly divide the world's thinkers into those who have and haven't undermined Western ethics, here once again the ultimate source of anti-Darwinian feeling may be moral gloom.

The case of *Commentary* looks more significant, however, because the magazine is published by the American Jewish Committee and is much concerned with defending Jewish beliefs and affinities. In lending their imprimatur to intelligent design, the editors can hardly have been unaware that they were joining forces with Christian zealots like Johnson, who has declared the Incarnation of Christ to be as certain as the proposition "that apples fall down rather than up," or like William Dembski, whose ultimate thesis is that "all disciplines find their completion in Christ and cannot be properly understood apart from Christ." But *Commentary*'s willingness to submerge religious differences for the sake of an imagined solidarity is nothing new. Rallying around both "family values" and the modern state that occupies the biblical Holy Land, the magazine's guiding figures had previously acknowledged that they share some principles with the evangelical right. That realignment reached a memorable climax when, in 1995, Norman Podhoretz extended a friendly hand to Pat Robertson despite the latter's authorship of *The New World Order,* a *Protocols*-style tract against "the Jews."

Commentary prides itself on favoring pragmatic realism over wishful thinking; and where science and technology are concerned, you can expect its articles to claim the support of authenticated research. But there is one exception: evolutionary biology has been consigned to the Johnsonian limbo of "materialistic philosophy." Such, among those who see themselves as guardians of decency and order, is the power of resistance to the disturbing prospect of a world unsupervised by a transcendent moral sovereign. The result is that *Commentary*, in the company of other magazines that treat natural selection as an illusion, tacitly encourages creationists to advance toward their primary goal: adulterating the public school curriculum so that children and adolescents will be denied access to an empirically plausible understanding of human origins.

But what about the secular left? Surely, one might suppose, that faction, with its reflexive aversion to "faith-based" initiatives, can be counted upon to come to the aid of embattled evolutionism, and doubly so when some of the attacks are mounted in organs like *Commentary* and *The New Criterion*. This expectation, however, overlooks the antiscientific bias that has characterized much leftist thought for the past quarter century.

Liberals and radicals who have been taught in college to believe that rival scientific paradigms are objectively incommensurable, that the real arbiter between theories is always sociopolitical power, and that Western science has been an oppressor of dispossessed women, minorities, and workers will be lukewarm at best toward Darwin. The latter, after all, shared the prejudices of his age and allowed some of them to inform his speculations about racial hierarchy and innate female character. Then, too, there is the sorry record of Social Darwinism to reckon with. Insofar as it has become habitual to weigh theories according to the attitudinal failings of their devisers and apostles, natural selection is shunned by some progressives, who are thus in no position to resist the creationist offensive. And while other leftists do broadly accede to evolutionism, much of their polemical energy is directed not against creationists but against Darwinian "evolutionary psychologists," a.k.a. sociobiologists, who speculate about the adaptive origins of traits and institutions that persist today.

Political suspicion on the left; fear of chaos on the right. Who will stand up for evolutionary biology and insist that it be taught without censorship or dilution? And who will register its challenge to human vanity without flinching? The answer seems obvious at first: people who employ Darwinian theory in their professional work. But even in this group we will see that frankness is less common than waffling and confusion. The problem, once again, is how to make room for God.

As even Phillip Johnson concedes, most of our religious sects are formally opposed to the campaign against Darwinism. Various church councils have avowed that evolution poses no threat to supernatural belief, and the same position is eagerly endorsed by scientific bodies. Creationists who read those declarations, however, always notice that a key question has been fudged. What *kind* of God is consistent with evolutionary theory? Theistic evolutionism would seem to demote the shaper of the universe to a *deus absconditus* who long ago set some processes in motion and then withdrew from the scene. And we have already noted that even this faint whiff of divinity is more than the theory of natural selection strictly requires.

Because Americans on the whole profess faith in both science and a personal God, those who experience this conflict are eager to be told that it is easily resolved. The public appetite for such reassurance is never sated. Not surprisingly, then, universities have developed specialties in "science and religion," and one book of soothing wisdom can hardly be scanned before the next entry appears in print. When coldly examined, however, these productions almost invariably prove to have adulterated scientific doctrine or to have emptied religious dogma of its commonly accepted meaning. And this legerdemain is never more brazen than when the scientific topic is Darwinism.

Take, for example, *The Faith of Biology and the Biology of Faith* by Robert Pollack, a molecular biologist at Columbia University and the director of its recently founded Center for the Study of Science and Religion. The title of Pollack's book appears to promise a vision encompassing the heavens above and the lab below. By the time he gets to evolution on page 2, however, the project has already collapsed. There he tells us that a Darwinian understanding of the natural world "is simply too terrifying and depressing to me to be borne without the emotional buffer of my own religion." By cleaving to the Torah he can lend "an irrational certainty of meaning and purpose to a set of data that otherwise show no sign of supporting any meaning to our lives on earth beyond that of being numbers in a cosmic lottery with no paymaster."

If Pollack's argument had stopped at this point, he could at least be praised for candor about his failure of nerve. But he is determined to place "feelings on a par with facts," and his book is therefore studded with clumsy attempts to make religion and science coincide after all by means of word magic. The rabbi and the molecular biologist, he extravagantly proposes, "share two beliefs founded entirely on faith . . . : that one day the text of their choice will be completely understood and that on that day death will have no power over us." Moreover, he declares that scientific insight comes from "an intrinsically unknowable place" — and who is the Unknowable One, he asks, if not God himself? Hence there is "only a semantic difference between scientific insight and what is called, in religious terms, revelation."

Pollack's half-formed ideas and bumbling prose stand in sharp contrast to the suaveness of John F. Haught, a professor of theology and director of yet another Center for the Study of Science and Religion, this one at Georgetown University. In *God After Darwin: A Theology of Evolution,* Haught acknowledges that the cruel indifference of Darwinian nature ought to jar the complacency of his fellow Catholics. But Haught himself remains unruffled; we need only bear in mind, he says, that our "thoughts about God after Darwin must be continuous with the authoritative scriptural and traditional sources of faith." In that spirit, Haught blithely assimilates Darwin to "humility theology," a body of thought depicting a God who "participates fully in the world's struggle and pain" and who chose to make himself vulnerable through the Incarnation and Crucifixion.

What God wants from planet earth, Haught informs us, is "the building of 'soul' in humans." That job requires plenty of agony and death — just what we find, happily, in "the cruciform visage of nature reflected in Darwinian science." Evolution occurs, then, "because God is more interested in adventure than in preserving the status quo." And though the story of emergence and extinction may look rather drawn-out and impersonal from our sublunary point of view, Haught assures us that it's all going to be redeemed: "Everything whatsoever that occurs in evolution — all the suffering and tragedy as well as the emergence of new life and intense beauty — is 'saved' by being taken eternally into God's own feeling of the world."

Not surprisingly, Haught's favorite scientific figure is the long-discredited paleontologist and Jesuit priest Pierre Teilhard de Chardin, who argued, somewhat in Haught's own lofty style, that the evolutionary process is being drawn forward to an "Omega point," a universal acceptance of Jesus Christ as Lord. As Haught remarks, this conception relocates God in the future and depicts him not as a planner but as "a transcendent force of attraction." But it doesn't occur to Haught that such teleology is just what Darwin managed to subtract from science. Whether pushing us or pulling us toward his desired end, the Christian God is utterly extraneous to evolution as Darwin and his modern successors have understood it. Evolution is an undirected, reactive process — the exact opposite of Haught's construal — or it is nothing at all.

Unlike Pollack and Haught, the philosopher of science Michael Ruse, who now teaches at Florida State University, has an expert's understanding of Darwinian theory and a creditable history of standing up to creationists in court testimony. That experience has doubtless shown him how advantageous it is, in God's country, for proponents of evolution to earn the support of religious believers. In *Can a Darwinian Be a Christian?* the agnostic Ruse contends that only "tensions rather than absolute and ineradicable contradictions" subsist between evolutionary and Christian doctrine, and he is sure that those tensions can be eased if both parties resolve to ponder "where and how they might be prepared to compromise. . . ." But what gets compromised when Ruse attempts to build this conciliatory case is his own fidelity to the essential features of Darwinism.

Consider the question of original sin. An evolutionist, Ruse concedes, cannot be expected to lend credence to a guilt literally inherited from Adam's primordial transgression. Nevertheless, if we are willing to interpret the concept liberally, "a ready understanding of original sin offers itself." Successful adaptations generally "involve self-interest, if not outright selfishness, with the host of features and attitudes and characteristics that we all find offensive and that the Christian judges sinful." Hence "original sin is part of the biological package."

Can Ruse be serious here? How could the result of fortunate mutations be called a sin? Both "selfishness" and "altruism" are found in nature for the same amoral reason: under given circumstances they yield an adaptive (and eventually reproductive) advantage. Ruse has blundered into gross anthropomorphism, ascribing psychological and moral traits to organisms that were programmed by natural selection to attack, poison, and deceive without cogitation.

And then there is the Darwinian/Christian impasse over miracles. According to Ruse, we needn't suppose that Jesus actually walked on water, produced food from nowhere, and raised the dead. Perhaps "people's hearts were so filled with love by Jesus' talk and presence that . . . they shared" their loaves and fishes. Or again, Lazarus may have been in a trance when awakened, and so perhaps was Jesus himself before his alleged resurrection — which, on the other hand, may have been only a metaphor for the "great joy and hope" excited in his followers. On and on plods Ruse, cheerfully turning wonders into banalities. Inside every Baptist, he seems to believe, there is a Unitarian struggling to get free.

In this book, but nowhere else in his soberly rational works, Ruse treats propositions about "God's judgment, the appropriateness of His righteous punishment, and the need and meaningfulness of His grace and forgiveness" as if they carried the same epistemic weight as propositions about the ancestry of birds and dolphins. As he knows full well, it just isn't so. However acrimoniously scientists may quarrel, they subscribe to canons of evidence that refer every dispute to the arbitration of discovered facts, whereas divine judgment, punishment, grace, and forgiveness are *irresolvably* mysterious. Insofar as Ruse tries to put that difference under the rug, he forsakes the empirical tradition that he has elsewhere worked so hard to protect.

The most startling disjunction of sensibility, however — a Jekyll–Hyde metamorphosis between the covers of one book — is manifested in Kenneth Miller's *Finding Darwin's God: A Scientist's Search for Common Ground Between God and Evolution,* a work whose first half, as I suggested, constitutes the most trenchant refutation of the newer creationism to be found anywhere. Yet when Miller then tries to drag God and Darwin to the bargaining table, his sense of proportion and probability abandons him, and he himself proves to be just another "God of the gaps" creationist. That is, he joins Phillip Johnson, William Dembski, and company in seizing upon the not-yet-explained as if it must be a locus of intentional action by the Christian deity.

Like the sophists of intelligent design, Miller rounds up the usual atheistic suspects — Daniel Dennett and Richard Dawkins, along with Cornell's William B. Provine and Harvard's Edward O. Wilson and Richard Lewontin — and represents them as dangling before us Satan's offer: "Exchange your belief in God for a material theology of disbelief, and complete knowledge will be yours." As always, the choice is stark: we must either surrender to such meretricious temptation or leave some sensible room for theology of the more familiar kind. With Michael Behe and John Haught, Miller wants his biology to sit comfortably with the dogmas of Roman Catholicism — for example, that Jesus was born of a virgin. Such a contention, he says, "makes no scientific sense," but that's just the point. "What can science say about a miracle? Nothing. By definition, the miraculous is beyond explanation, beyond our understanding, beyond science."

As Miller realizes, however, an appeal to the ineffable contributes nothing to a project called "finding Darwin's God." His only recourse, if he is to stay faithful to Darwinian theory, is to make a more modest case for a measure of unpredictability that can then be given a theological spin. "What if the regularities of nature," he asks, "were fashioned in such a way that they *themselves* allowed for the divine?" Quantum indeterminacy must have allowed God to shape evolution on the subatomic level "with care and with subtlety," gently nudging matter toward the emergence of "exactly what He was looking for — a creature who, like us, could know Him and love Him, could perceive the heavens and dream of the stars, a creature who would eventually discover the extraordinary process of evolution that filled His earth with so much life."

This case differs only marginally from the intelligent design argument that Miller decisively refuted in the opening chapters of his book. The distinction is simply that Miller's Darwinian God wouldn't have known in advance that you and I, who have finally pleased him by tumbling to his evolutionary scheme, would emerge from a line of apes. "Theologically," Miller explains, "the care that God takes *not* to intervene pointlessly in the world is an essential part of His plan for us" — or rather, of his plan for some intelligent species that luckily turned out to be us. Now that we're here, though, we humans can regard ourselves as "*both* the products of evolution and the apple of God's eye."

"In each age," Miller writes, God "finds a way to bring His message directly to us." But which divine message, among earth's thousands, does he mean? Although he notes in passing that the Almighty neglected to get his redemptive word out to the Mayas and the Toltecs among others, he dismisses that anomaly with an indifferent shrug. By effectively reducing religion to the Western monotheisms and then glossing over *their* differences, he blots from view the world's pantheist gurus, animist shamans, and idol worshipers while making the quarrelsome ayatollahs, cardinals, presbyters, and rabbis look as if they are hearing the same clear voice from above.

Miller doesn't explain how he has been able to delve so unerringly into the Architect's cravings, schemes, and limitations. Nor does he answer the question that he himself crushingly deployed against the ID team: "Why did this magician, in order to produce the contemporary world, find it necessary to create and destroy creatures, habitats, and ecosystems millions of times over?" The God who entrusted his will entirely to mutation and selection can hardly be the one who, as Miller alleges, presented the ancient Hebrews with an ethical guidebook, "knowing exactly what they would understand"; who transformed himself into a man so as to settle accounts in his ledger of human sin; who "has a plan for each of us"; and who has endowed us with "immortal souls." As the fruit of a keen scientific mind, *Finding Darwin's God* appears to offer the strongest corroboration yet of William Provine's infamous rule: if you want to marry Christian doctrine with modern evolutionary biology, "you have to check your brains at the church-house door."

There is, however, one last way of ensuring that Darwinism won't inhibit religious belief and vice versa. It is proposed by America's best-known paleontologist, Stephen Jay Gould, whose record of opposition to creationism and to religious interference with scientific research is consistent and unimpeachable. In *Rocks of Ages: Science and Religion in the Fullness of Life,* Gould maintains that the two "magisteria," or domains of authority, will enjoy mutual respect if their adherents refrain from any attempted synthesis. With "NOMA" — that's "nonoverlapping magisteria" — kept firmly in mind, scientists and divines can carry out their equally valuable tasks, the investigation of nature and the pursuit of spiritual values and ethical rules, without trespassing on one another's terrain.

Gould's term "magisteria" was inspired by two popes who have issued dictates about evolution. In *Humani Generis* (1950), Pius XII ruled physical evolution to be compatible with orthodox faith but still unproven, and he warned against any supposition that the soul had emerged from natural processes. And in 1996 John Paul II took note of the convergent findings that by then had rendered evolution "more than a hypothesis" — a conclusion that Gould hails as his "favorite example of NOMA" emanating from an unexpected religious source. If this is really the pope's considered view, says Gould, "we may rejoice in a pervasive and welcome consensus" between scientists and ecclesiastics.

Regrettably, however, Gould barely hints at a crucial point that ought to have muted his hosanna. John Paul II's position on the supernatural origin of the soul is identical to that of every predecessor pope. "The Church's Magisterium," he wrote in the very statement that Gould hails,

> is directly concerned with the question of evolution, for it involves the conception of man: Revelation teaches that he was created in the image and likeness of God (cf. Gn 1:27–29). The conciliar Constitution *Gaudium et spes* has magnificently explained this doctrine, which . . . recalled that man is "the only creature on earth that God has wanted for its own sake." . . . Pius XII stressed this essential point: if the human body takes its origin from pre-existent living matter, the spiritual soul is immediately created by God. . . .
>
> Consequently, theories of evolution which, in accordance with the philosophies inspiring them, consider the mind as emerging from the forces of living matter, or as a mere epiphenomenon of this matter, are incompatible with the truth about man.

This passage shows that the Church, while conceding that evolutionary science can no longer be snubbed, remains intransigently creationist where its own interests are concerned. Nor has Gould been unmindful of that fact. When he broached the NOMA rule in his *Natural History* column of March 1997, he voiced a suspicion that John Paul II's "insistence on divine infusion of the soul" was "a device for maintaining a belief in human superiority within an evolutionary world offering no privileged position to any creature." But he backed down at once, pleading in his next sentence that "souls represent a subject outside the magisterium of science." And now in *Rocks of Ages,* borrowing heavily from his *Natural History* piece, he has chosen to omit any mention of his misgivings.

Gould's concordat sounds more reasonable than the pope's until one asks what it might mean in practice. As a paleontologist who was raised without a faith, Gould could be expected to feel more protective of one magisterium than the other. Sure enough, his NOMA forbids the miraculous and, by extension, any idea of divine action within the world: "Thou shalt not mix the magisteria by claiming that God directly ordains important events in the history of nature by special interference knowable only through revelation and not accessible to science." As Phillip Johnson has understandably complained, "This is 'separate but equal' of the *apartheid* variety." And John Haught chimes in, "No conceivable theology, by definition, could ever live comfortably with evolution if Gould's claim is correct that Darwin's theory inevitably entails a cosmos devoid of directionality and overall significance."

Rocks of Ages sounds like an extended effort to soften Gould's peace-on-my-terms through flattery of the pious. The strain is apparent in his uncharacteristically slick and sentimental prose:

> . . . Science gets the age of rocks, and religion the rock of ages; science studies how the heavens go, religion how to go to heaven.

> I join nearly all people of goodwill in wishing to see two old and cherished institutions, our two rocks of ages — science and religion — coexisting in peace while each works to make a distinctive patch for the integrated coat of many colors that will celebrate the distinctions of our lives, yet cloak human nakedness in a seamless covering called wisdom.

Compare these excerpts, for tone and content, with what Gould spontaneously told a television interviewer in 1998:

> I think that notion that we are all in the bosom of Abraham or are in God's embracing love is — look, it's a tough life and if you can delude

> yourself into thinking that there's all some warm and fuzzy meaning to it all, it's enormously comforting. But I do think it's just a story we tell ourselves.

I am not the first commentator to point out that there is something contradictory about Gould's attempt, in *Rocks of Ages,* to endear himself to believers by praising their "wisdom" while reprimanding atheists for their "aggressive advocacy" of a position scarcely distinguishable from his own.

On one point, however, Gould is perfectly candid. In order to make his "nonoverlapping magisteria" palatable to both parties, he has set aside considerations of truth and followed what he calls a "'Goldilocks principle' of 'just right' between too much and too little. . . . NOMA represents the bed of proper firmness, and the right amount of oatmeal at the right temperature." "Oatmeal" is right. Instead of grappling with the issues that seriously divide religious and scientific thinkers, as Ruse and Miller at least attempt to do, Gould delivers gratuitous restraining orders to both factions. In exchange for abandoning their immanent God and settling for a watery deism, the religionists get the realm of ethics largely to themselves, while scientists are admonished to eschew "invalid forays into the magisterium of moral argument." But *Rocks of Ages* is itself a moral argument proffered by a scientist and an infidel — and why not?

As Gould maintains, scientific facts and theories don't tell us how we ought to conduct ourselves. This doesn't mean, however, that ethics can be confidently entrusted to shepherds of souls. Undoubtedly, fear of God makes for social cohesion and moral restraint, at least toward those who share our faith; and many noble causes are championed by people who think they are implementing his wishes. But religious certitude can also remain fixated on ancient prejudices and prohibitions that dehumanize outsiders, coarsen ethical calculation, and retard social enlightenment. If he weren't bent on playing the roving ambassador between two wary camps, Gould would be the first person to acknowledge this obvious truth.

Both Gould and many of his readers want to believe that liberty begins where biology ceases to hold sway. We live, he writes, in a universe "indifferent to our suffering, and therefore offering us maximal freedom to thrive, or to fail, in our own chosen way." Ringing words, but what do they mean? The universe is also indifferent to a mouse being tossed and tortured by a cat, but the mouse's freedom is no greater for that. The options we enjoy as a species that has staked its fate on intelligence and foresight are surely a gift of our staggeringly complex neural circuitry, which is natural selection's boldest experiment in trading blind instinct for feedback mechanisms that allow dangers to be consciously assessed and circumvented. By shifting levels of discourse and proclaiming that we acquire our scope for action from a mere absence of interference by "the universe," Gould has momentarily left science behind and become a theologian, albeit an existentialist one.

The evasions practiced by Pollack, Haught, Ruse, Miller, and now Gould, in concert with those of the intelligent design crew, remind us that Darwinism, despite its radical effect on science, has yet to temper the self-centered way in which we assess our place and actions in the world. Think of the shadows now falling across our planet: overpopulation, pollution, dwindling and maldistributed resources, climatic disruption, new and resurgent plagues, ethnic and religious hatred, the ravaging of forests and jungles, and the consequent loss of thousands of species per year — the greatest mass extinction, it has been said, since the age of the dinosaurs. So long as we regard ourselves as creatures apart who need only repent of our personal sins to retain heaven's blessing, we won't take the full measure of our species-wide responsibility for these calamities.

An evolutionary perspective, by contrast, can trace our present woes to the dawn of agriculture ten thousand years ago, when, as Niles Eldredge has observed, we became "the first species in the entire 3.8-billion-year history of life to stop living inside local ecosystems." Today, when we have burst from six million to six billion exploiters of a biosphere whose resilience can no longer be assumed, the time has run out for telling ourselves that we are the darlings of a deity who placed nature here for our convenience. We are the most resourceful, but also the most dangerous and disruptive, animals in this corner of the universe. A Darwinian understanding of how we got that way could be the first step toward a wider ethics commensurate with our real transgressions, not against God but against earth itself and its myriad forms of life.

BARBARA EHRENREICH

Welcome to Cancerland

FROM *Harper's Magazine*

I WAS THINKING of it as one of those drive-by mammograms, one stop in a series of mundane missions including post office, supermarket, and gym, but I began to lose my nerve in the changing room, and not only because of the kinky necessity of baring my breasts and affixing tiny X-ray opaque stars to the tip of each nipple. I had been in this place only four months earlier, but that visit was just part of the routine cancer surveillance all good citizens of HMOs or health plans are expected to submit to once they reach the age of fifty, and I hadn't really been paying attention then. The results of that earlier session had aroused some "concern" on the part of the radiologist and her confederate, the gynecologist, so I am back now in the role of a suspect, eager to clear my name, alert to medical missteps and unfair allegations. But the changing room, really just a closet off the stark windowless space that houses the mammogram machine, contains something far worse, I notice for the first time now — an assumption about who I am, where I am going, and what I will need when I get there. Almost all of the eye-level space has been filled with photocopied bits of cuteness and sentimentality: pink ribbons, a cartoon about a woman with iatrogenically flattened breasts, an "Ode to a Mammogram," a list of the "Top Ten Things Only Women Understand" ("Fat Clothes" and "Eyelash Curlers" among them), and, inescapably, right next to the door, the poem "I Said a Prayer for You Today," illustrated with pink roses.

It goes on and on, this mother of all mammograms, cutting into gym time, dinnertime, and lifetime generally. Sometimes the machine doesn't work, and I get squished into position to no purpose at all. More often, the X ray is successful but apparently alarming to the invisible radiologist, off in some remote office, who calls the shots and never has the courtesy to show her face with an apology or an explanation. I try pleading with the technician: I have no known risk factors, no breast cancer in the family, had my babies relatively young and nursed them both. I eat right, drink sparingly, work out, and doesn't that count for something? But she just gets this tight little professional smile on her face, either out of guilt for the torture she's inflicting or because she already knows something that I am going to be sorry to find out for myself. For an hour and a half the procedure is repeated: the squishing, the snapshot, the technician bustling off to consult the radiologist and returning with a demand for new angles and more definitive images. In the intervals while she's off with the doctor I read the *New York Times* right down to the personally irrelevant sections like theater and real estate, eschewing the stack of women's magazines provided for me, much as I ordinarily enjoy a quick read about sweat-proof eyeliners and "fabulous sex tonight," because I have picked up this warning vibe in the changing room, which, in my increasingly anxious state, translates into: femininity is death. Finally there is nothing left to read but one of the free local weekly newspapers, where I find, buried deep in the classifieds, something even more unsettling than the growing prospect of major disease — a classified ad for a "breast cancer teddy bear" with a pink ribbon stitched to its chest.

Yes, atheists pray in their foxholes — in this case, with a yearning new to me and sharp as lust, for a clean and honorable death by shark bite, lightning strike, sniper fire, car crash. Let me be hacked to death by a madman, is my silent supplication — anything but suffocation by the pink sticky sentiment embodied in that bear and oozing from the walls of the changing room.

My official induction into breast cancer comes about ten days later with the biopsy, which, for reasons I cannot ferret out of the surgeon, has to be a surgical one, performed on an outpatient basis but under general anesthesia, from which I awake to find him standing perpendicular to me, at the far end of the gurney, down near my feet, stating gravely, "Unfortunately, there is a cancer." It takes me all the rest of that drug-addled day to decide that the most heinous thing about that sentence is not the presence of cancer but the absence of me — for I, Barbara, do not enter into it even as a location, a geographical reference point. Where I once was — not a commanding presence perhaps but nonetheless a standard assemblage of flesh and words and gesture — "there is a cancer." I have been replaced by it, is the surgeon's implication. This is what I am now, medically speaking.

In my last act of dignified self-assertion, I request to see the pathology slides myself. This is not difficult to arrange in our small-town hospital, where the pathologist turns out to be a friend of a friend, and my rusty Ph.D. in cell biology (Rockefeller University, 1968) probably helps. He's a jolly fellow, the pathologist, who calls me "hon" and sits me down at one end of the dual-head microscope while he mans the other and moves a pointer through the field. These are the cancer cells, he says, showing up blue because of their overactive DNA. Most of them are arranged in staid semicircular arrays, like suburban houses squeezed into a cul-de-sac, but I also see what I know enough to know I do not want to see: the characteristic "Indian files" of cells on the march. The "enemy," I am supposed to think — an image to save up for future exercises in "visualization" of their violent deaths at the hands of the body's killer cells, the lymphocytes and macrophages. But I am impressed, against all rational self-interest, by the energy of these cellular conga lines, their determination to move on out from the backwater of the breast to colonize lymph nodes, bone marrow, lungs, and brain. These are, after all, the fanatics of Barbaraness, the rebel cells that have realized that the genome they carry, the genetic essence of me, has no further chance of normal reproduction in the postmenopausal body we share, so why not just start multiplying like bunnies and hope for a chance to break out?

It has happened, after all; some genomes have achieved immortality through cancer. When I was a graduate student, I once asked about the strain of tissue-culture cells labeled "HeLa" in the heavy-doored room maintained at body temperature. "HeLa," it turns out, refers to one Henrietta Lacks, whose tumor was the progenitor of all HeLa cells. She died; they live, and will go on living until someone gets tired of them or forgets to change their tissue-culture medium and leaves them to starve. Maybe this is what my rebel cells have in mind, and I try beaming them a solemn warning: the chances of your surviving me in tissue culture are nil. Keep up this selfish rampage and you go down, every last one of you, along with the entire Barbara enterprise. But what kind of a role model am I, or are multicellular human organisms generally, for putting the common good above mad anarchistic individual ambition? There is a reason, it occurs to me, why cancer is our metaphor for so many runaway social processes, like corruption and "moral decay": we are no less out of control ourselves.

After the visit to the pathologist, my biological curiosity drops to a lifetime nadir. I know women who followed up their diagnoses with weeks or months of self-study, mastering their options, interviewing doctor after doctor, assessing the damage to be expected from the available treatments. But I can tell from a few hours of investigation that the career of a breast-cancer patient has been pretty well mapped out in advance for me: you may get to negotiate the choice between lumpectomy and mastectomy, but lumpectomy is commonly followed by weeks of radiation, and in either case if the lymph nodes turn out, upon dissection, to be invaded — or "involved," as it's less threateningly put — you're doomed to chemotherapy, meaning baldness, nausea, mouth sores, immunosuppression, and possible anemia. These interventions do not constitute a "cure" or anything close, which is why the death rate from breast cancer has changed very little since the 1930s, when mastectomy was the only treatment available. Chemotherapy, which became a routine part of breast-cancer treatment in the eighties, does not confer anywhere near as decisive an advantage as patients are often led to believe, especially in postmenopausal women like myself — a two or three percentage point difference in ten-year survival rates,* according to America's best-known breast-cancer surgeon, Dr. Susan Love.

I know these bleak facts, or sort of know them, but in the fog of anesthesia that hangs over those first few weeks, I seem to lose my capacity for self-defense. The pressure is on, from doctors and loved ones, to do something right away — kill it, get it out now. The endless exams, the bone scan to check for metastases, the high-

* In the United States, one in eight women will be diagnosed with breast cancer at some point. The chances of her surviving for five years are 86.8 percent. For a black woman this falls to 72 percent; and for a woman of any race whose cancer has spread to the lymph nodes, to 77.7 percent.

tech heart test to see if I'm strong enough to withstand chemotherapy — all these blur the line between selfhood and thing-hood anyway, organic and inorganic, me and it. As my cancer career unfolds, I will, the helpful pamphlets explain, become a composite of the living and the dead — an implant to replace the breast, a wig to replace the hair. And then what will I mean when I use the word *I*? I fall into a state of unreasoning passive aggressivity: they diagnosed this, so it's their baby. They found it, let them fix it.

I could take my chances with "alternative" treatments, of course, like punk novelist Kathy Acker, who succumbed to breast cancer in 1997 after a course of alternative therapies in Mexico, or actress and ThighMaster promoter Suzanne Somers, who made tabloid headlines last spring by injecting herself with mistletoe brew. Or I could choose to do nothing at all beyond mentally exhorting my immune system to exterminate the traitorous cellular faction. But I have never admired the "natural" or believed in the "wisdom of the body." Death is as "natural" as anything gets, and the body has always seemed to me like a retarded Siamese twin dragging along behind me, an hysteric really, dangerously overreacting, in my case, to everyday allergens and minute ingestions of sugar. I will put my faith in science, even if this means that the dumb old body is about to be transmogrified into an evil clown — puking, trembling, swelling, surrendering significant parts, and oozing postsurgical fluids. The surgeon — a more genial and forthcoming one this time — can fit me in; the oncologist will see me. Welcome to Cancerland.

Fortunately, no one has to go through this alone. Thirty years ago, before Betty Ford, Rose Kushner, Betty Rollin, and other pioneer patients spoke out, breast cancer was a dread secret, endured in silence and euphemized in obituaries as a "long illness." Something about the conjuncture of "breast," signifying sexuality and nurturance, and that other word, suggesting the claws of a devouring crustacean, spooked almost everyone. Today, however, it's the biggest disease on the cultural map, bigger than AIDS, cystic fibrosis, or spinal injury, bigger even than those more prolific killers of women — heart disease, lung cancer, and stroke. There are roughly hundreds of Web sites devoted to it, not to mention newsletters, support groups, a whole genre of first-person breast-cancer books; even a glossy, upper-middle-brow, monthly magazine,

Mamm. There are four major national breast-cancer organizations, of which the mightiest, in financial terms, is the Susan G. Komen Foundation, headed by breast-cancer veteran and Bush's nominee for ambassador to Hungary Nancy Brinker. Komen organizes the annual Race for the Cure©, which attracts about a million people — mostly survivors, friends, and family members. Its Web site provides a microcosm of the new breast-cancer culture, offering news of the races, message boards for accounts of individuals' struggles with the disease, and a "marketplace" of breast-cancer-related products to buy.

More so than in the case of any other disease, breast-cancer organizations and events feed on a generous flow of corporate support. Nancy Brinker relates how her early attempts to attract corporate interest in promoting breast cancer "awareness" were met with rebuff. A bra manufacturer, importuned to affix a mammogram-reminder tag to his product, more or less wrinkled his nose. Now breast cancer has blossomed from wallflower to the most popular girl at the corporate charity prom. While AIDS goes begging and low-rent diseases like tuberculosis have no friends at all, breast cancer has been able to count on Revlon, Avon, Ford, Tiffany, Pier 1, Estée Lauder, Ralph Lauren, Lee Jeans, Saks Fifth Avenue, JC Penney, Boston Market, Wilson athletic gear — and I apologize to those I've omitted. You can "shop for the cure" during the week when Saks donates 2 percent of sales to a breast-cancer fund; "wear denim for the cure" during Lee National Denim Day, when for a five-dollar donation you get to wear blue jeans to work. You can even "invest for the cure," in the Kinetics Assets Management's new no-load Medical Fund, which specializes entirely in businesses involved in cancer research.

If you can't run, bike, or climb a mountain for the cure — all of which endeavors are routine beneficiaries of corporate sponsorship — you can always purchase one of the many products with a breast-cancer theme. There are 2.2 million American women in various stages of their breast-cancer careers, who, along with anxious relatives, make up a significant market for all things breast-cancer-related. Bears, for example: I have identified four distinct lines, or species, of these creatures, including "Carol," the Remembrance Bear; "Hope," the Breast Cancer Research Bear, which wears a pink turban as if to conceal chemotherapy-induced baldness; the "Susan Bear," named for Nancy Brinker's deceased sister, Susan; and the new Nick & Nora Wish Upon a Star Bear, available, along with the Susan Bear, at the Komen Foundation Web site's "marketplace."

And bears are only the tip, so to speak, of the cornucopia of pink-ribbon-themed breast-cancer products. You can dress in pink-beribboned sweatshirts, denim shirts, pajamas, lingerie, aprons, loungewear, shoelaces, and socks; accessorize with pink rhinestone brooches, angel pins, scarves, caps, earrings, and bracelets; brighten up your home with breast-cancer candles, stained-glass pink-ribbon candleholders, coffee mugs, pendants, wind chimes, and night-lights; pay your bills with special BreastChecks or a separate line of Checks for the Cure. "Awareness" beats secrecy and stigma of course, but I can't help noticing that the existential space in which a friend has earnestly advised me to "confront [my] mortality" bears a striking resemblance to the mall.

This is not, I should point out, a case of cynical merchants exploiting the sick. Some of the breast-cancer tchotchkes and accessories are made by breast-cancer survivors themselves, such as "Janice," creator of the "Daisy Awareness Necklace," among other things, and in most cases a portion of the sales goes to breast-cancer research. Virginia Davis of Aurora, Colorado, was inspired to create the "Remembrance Bear" by a friend's double mastectomy and sees her work as more of a "crusade" than a business. This year she expects to ship ten thousand of these teddies, which are manufactured in China, and send part of the money to the Race for the Cure. If the bears are infantilizing — as I try ever so tactfully to suggest is how they may, in rare cases, be perceived — so far no one has complained. "I just get love letters," she tells me, "from people who say, 'God bless you for thinking of us.'"

The ultrafeminine theme of the breast-cancer "marketplace" — the prominence, for example, of cosmetics and jewelry — could be understood as a response to the treatments' disastrous effects on one's looks. But the infantilizing trope is a little harder to account for, and teddy bears are not its only manifestation. A tote bag distributed to breast cancer patients by the Libby Ross Foundation (through places such as the Columbia Presbyterian Medical Center) contains, among other items, a tube of Estée Lauder Perfumed Body Crème, a hot-pink satin pillowcase, an audiotape "Meditation to Help You with Chemotherapy," a small tin of peppermint pastilles, a set of three small inexpensive rhinestone bracelets, a pink-striped "journal and sketch book," and — somewhat jarringly — a small box of crayons. Marla Willner, one of the founders of the Libby Ross Foundation, told me that the crayons "go with the journal — for people to express different moods, different thoughts . . ." though she admitted she has never tried to write with crayons herself. Possibly the idea is that regression to a state of childlike dependency puts one in the best frame of mind with which to endure the prolonged and toxic treatments. Or it may be that, in some versions of the prevailing gender ideology, femininity is by its nature incompatible with full adulthood — a state of arrested development. Certainly men diagnosed with prostate cancer do not receive gifts of Matchbox cars.

But I, no less than the bear huggers, need whatever help I can get, and start wading out into the Web in search of practical tips on hair loss, lumpectomy versus mastectomy, how to select a chemotherapy regimen, what to wear after surgery and eat when the scent of food sucks. There is, I soon find, far more than I can usefully absorb, for thousands of the afflicted have posted their stories, beginning with the lump or bad mammogram, proceeding through the agony of the treatments; pausing to mention the sustaining forces of family, humor, and religion; and ending, in almost all cases, with warm words of encouragement for the neophyte. Some of these are no more than a paragraph long — brief waves from sister sufferers; others offer almost hour-by-hour logs of breast-deprived, chemotherapized lives:

> Tuesday, August 15, 2000: Well, I survived my 4th chemo. Very, very dizzy today. Very nauseated, but no barfing! It's a first. . . . I break out in a cold sweat and my heart pounds if I stay up longer than 5 minutes.

> Friday, August 18, 2000: . . . By dinner time, I was full out nauseated. I took some meds and ate a rice and vegetable bowl from Trader Joe's. It smelled and tasted awful to me, but I ate it anyway. . . . Rick brought home some Kern's nectars and I'm drinking that. Seems to have settled my stomach a little bit.

I can't seem to get enough of these tales, reading on with panicky fascination about everything that can go wrong — septicemia, ruptured implants, startling recurrences a few years after the completion of treatments, "mets" (metastases) to vital organs, and — what scares me most in the short term — "chemo-brain," or the cognitive deterioration that sometimes accompanies chemotherapy. I compare myself with everyone, selfishly impatient with those whose conditions are less menacing, shivering over those who have reached Stage IV ("There is no Stage V," as the main character in *Wit,* who has ovarian cancer, explains), constantly assessing my chances.

Feminism helped make the spreading breast-cancer sisterhood possible, and this realization gives me a faint feeling of belonging. Thirty years ago, when the disease went hidden behind euphemism and prostheses, medicine was a solid patriarchy, women's bodies its passive objects of labor. The Women's Health Movement, in which I was an activist in the seventies and eighties, legitimized self-help and mutual support and encouraged women to network directly, sharing their stories, questioning the doctors, banding together. It is hard now to recall how revolutionary these activities once seemed, and probably few participants in breast-cancer chat rooms and message boards realize that when post-mastectomy patients first proposed meeting in support groups in the mid-1970s, the American Cancer Society responded with a firm and fatherly "no." Now no one leaves the hospital without a brochure directing her to local support groups and, at least in my case, a follow-up call from a social worker to see whether I am safely ensconced in one. This cheers me briefly, until I realize that if support groups have won the stamp of medical approval, this may be because they are no longer perceived as seditious.

In fact, aside from the dilute sisterhood of the cyber (and actual) support groups, there is nothing very feminist — in an ideological or activist sense — about the mainstream of breast-cancer culture today. Let me pause to qualify: you can, if you look hard enough, find plenty of genuine, self-identified feminists within the vast pink sea of the breast-cancer crusade, women who are militantly determined to "beat the epidemic" and insistent on more user-friendly approaches to treatment. It was feminist health activists who led the campaign, in the seventies and eighties, against the most savage form of breast-cancer surgery — the Halsted radical mastectomy, which removed chest muscle and lymph nodes as well as breast tissue and left women permanently disabled. It was the Women's Health Movement that put a halt to the surgical practice, common in the seventies, of proceeding directly from biopsy to mastectomy without ever rousing the patient from anesthesia. More recently, feminist advocacy groups such as the San Francisco–based Breast Cancer Action and the Cambridge-based Women's Community Cancer Project helped blow the whistle on "high-dose chemotherapy," in which the bone marrow was removed prior to otherwise lethal doses of chemotherapy and later replaced — to no good effect, as it turned out.

Like everyone else in the breast-cancer world, the feminists want a cure, but they even more ardently demand to know the cause or causes of the disease without which we will never have any means of prevention. "Bad" genes of the inherited variety are thought to account for fewer than 10 percent of breast cancers, and only 30 percent of women diagnosed with breast cancer have any known risk factor (such as delaying childbearing or the late onset of menopause) at all. Bad lifestyle choices like a fatty diet have, after brief popularity with the medical profession, been largely ruled out. Hence suspicion should focus on environmental carcinogens, the feminists argue, such as plastics, pesticides (DDT and PCBs, for example, though banned in this country, are still used in many Third World sources of the produce we eat), and the industrial runoff in our ground water. No carcinogen has been linked definitely to human breast cancer yet, but many have been found to cause the disease in mice, and the inexorable increase of the disease in industrialized nations — about one percent a year between the 1950s and the 1990s — further hints at environmental factors, as does the fact that women migrants to industrialized countries quickly develop the same breast-cancer rates as those who are native born. Their emphasis on possible ecological factors, which is not shared by groups such as Komen and the American Cancer Society, puts the feminist breast-cancer activists in league with other, frequently rambunctious, social movements — environmental and anticorporate.

But today theirs are discordant voices in a general chorus of sentimentality and good cheer; after all, breast cancer would hardly be the darling of corporate America if its complexion changed from pink to green. It is the very blandness of breast cancer, at least in mainstream perceptions, that makes it an attractive object of corporate charity and a way for companies to brand themselves friends of the middle-aged female market. With breast cancer, "there was no concern that you might actually turn off your audience because of the life style or sexual connotations that AIDS has," Amy Langer, director of the National Alliance of Breast Cancer Organizations, told the *New York Times* in 1996. "That gives corporations a certain freedom and a certain relief in supporting the cause." Or as Cindy Pearson, director of the National Women's Health Network, the organizational progeny of the Women's Health Movement, puts it more caustically: "Breast cancer provides a way of doing something for women, without being feminist."

In the mainstream of breast-cancer culture, one finds very little anger, no mention of possible environmental causes, few complaints about the fact that, in all but the more advanced, metastasized cases, it is the "treatments," not the disease, that cause illness and pain. The stance toward existing treatments is occasionally critical — in *Mamm,* for example — but more commonly grateful; the overall tone, almost universally upbeat. The Breast Friends Web site, for example, features a series of inspirational quotes: "Don't Cry Over Anything that Can't Cry Over You," "I Can't Stop the Birds of Sorrow from Circling my Head, But I Can Stop Them from Building a Nest in My Hair," "When Life Hands Out Lemons, Squeeze Out a Smile," "Don't wait for your ship to come in . . . Swim out to meet it," and much more of that ilk. Even in the relatively sophisticated *Mamm,* a columnist bemoans not cancer or chemotherapy but the end of chemotherapy, and humorously proposes to deal with her separation anxiety by pitching a tent outside her oncologist's office. So pervasive is the perkiness of the breast-cancer world that unhappiness requires a kind of apology, as when "Lucy," whose "long term prognosis is not good," starts her personal narrative on breastcancertalk.org by telling us that her story "is not the usual one, full of sweetness and hope, but true nevertheless."

There is, I discover, no single noun to describe a woman with breast cancer. As in the AIDS movement, upon which breast-cancer activism is partly modeled, the words *patient* and *victim,* with their aura of self-pity and passivity, have been ruled un-PC. Instead, we get verbs: those who are in the midst of their treatments are described as *battling* or *fighting,* sometimes intensified with *bravely* or *fiercely* — language suggestive of Katharine Hepburn with her face to the wind. Once the treatments are over, one achieves the status of *survivor,* which is how the women in my local support group identify themselves, AA-style, as we convene to share war stories and rejoice in our "survivorhood": "Hi, I'm Kathy and I'm a three-year survivor." For those who cease to be survivors and join the more than forty thousand American women who succumb to breast cancer each year — again, no noun applies. They are said to have "lost their battle" and may be memorialized by photographs carried at races for the cure — our lost, brave sisters, our fallen soldiers. But in the overwhelmingly Darwinian culture that has grown up around breast cancer, martyrs count for little; it is the "survivors" who merit constant honor and acclaim. They, after all, offer living proof that expensive and painful treatments may in some cases actually work.

Scared and medically weakened women can hardly be expected to transform their support groups into bands of activists and rush out into the streets, but the equanimity of breast-cancer culture goes beyond mere absence of anger to what looks, all too often, like a positive embrace of the disease. As "Mary" reports, on the Bosom Buds message board:

> I really believe I am a much more sensitive and thoughtful person now. It might sound funny but I was a real worrier before. Now I don't want to waste my energy on worrying. I enjoy life so much more now and in a lot of aspects I am much happier now.

Or this from "Andee":

> This was the hardest year of my life but also in many ways the most rewarding. I got rid of the baggage, made peace with my family, met many amazing people, learned to take very good care of my body so it will take care of me, and reprioritized my life.

Cindy Cherry, quoted in the *Washington Post,* goes further:

> If I had to do it over, would I want breast cancer? Absolutely. I'm not the same person I was, and I'm glad I'm not. Money doesn't matter anymore. I've met the most phenomenal people in my life through this. Your friends and family are what matter now.

The First Year of the Rest of Your Life, a collection of brief narratives with a foreword by Nancy Brinker and a share of the royalties going to the Komen Foundation, is filled with such testimonies to the redemptive powers of the disease: "I can honestly say I am happier now than I have ever been in my life — even before the breast cancer." "For me, breast cancer has provided a good kick in the rear to get me started rethinking my life. . . ." "I have come out stronger, with a new sense of priorities . . ." Never a complaint about lost time, shattered sexual confidence, or the long-term weakening of the arms caused by lymph-node dissection and radiation. What does not destroy you, to paraphrase Nietzsche, makes you a spunkier, more evolved sort of person.

The effect of this relentless brightsiding is to transform breast cancer into a rite of passage — not an injustice or a tragedy to rail against, but a normal marker in the life cycle, like menopause or graying hair. Everything in mainstream breast-cancer culture serves, no doubt inadvertently, to tame and normalize the disease: the diagnosis may be disastrous, but there are those cunning pink rhinestone angel pins to buy and races to train for. Even the heavy traffic in personal narratives and practical tips, which I found so useful, bears an implicit acceptance of the disease and the current barbarous approaches to its treatment: you can get so busy comparing attractive head scarves that you forget to question a form of treatment that temporarily renders you both bald and immuno-incompetent. Understood as a rite of passage, breast cancer resembles the initiation rites so exhaustively studied by Mircea Eliade: first there is the selection of the initiates — by age in the tribal situation, by mammogram or palpation here. Then come the requisite ordeals — scarification or circumcision within traditional cultures, surgery and chemotherapy for the cancer patient. Finally, the initiate emerges into a new and higher status — an adult and a warrior — or in the case of breast cancer, a "survivor."

And in our implacably optimistic breast-cancer culture, the disease offers more than the intangible benefits of spiritual upward mobility. You can defy the inevitable disfigurements and come out, on the survivor side, actually prettier, sexier, more femme. In the lore of the disease — shared with me by oncology nurses as well as by survivors — chemotherapy smoothes and tightens the skin, helps you lose weight; and, when your hair comes back, it will be fuller, softer, easier to control, and perhaps a surprising new color. These may be myths, but for those willing to get with the prevailing program, opportunities for self-improvement abound. The American Cancer Society offers the "Look Good . . . Feel Better" program, "dedicated to teaching women cancer patients beauty techniques to help restore their appearance and self-image during cancer treatment." Thirty thousand women participate a year, each copping a free makeover and bag of makeup donated by the Cosmetic, Toiletry, and Fragrance Association, the trade association of the cosmetics industry. As for that lost breast: after reconstruction, why not bring the other one up to speed? Of the more than fifty thousand mastectomy patients who opt for reconstruction each year, 17 percent go on, often at the urging of their plastic surgeons, to get additional surgery so that the remaining breast will "match" the more erect and perhaps larger new structure on the other side.

Not everyone goes for cosmetic deceptions, and the question of wigs versus baldness, reconstruction versus undisguised scar, defines one of the few real disagreements in breast-cancer culture. On the more avant-garde, upper-middle-class side, *Mamm* magazine — which features literary critic Eve Kosofsky Sedgwick as a columnist — tends to favor the "natural" look. Here, mastectomy scars can be "sexy" and baldness something to celebrate. The January 2001 cover story features women who "looked upon their baldness not just as a loss, but also as an opportunity: to indulge their playful sides . . . to come in contact, in new ways, with their truest selves." One decorates her scalp with temporary tattoos of peace signs, panthers, and frogs; another expresses herself with a shocking purple wig; a third reports that unadorned baldness makes her feel "sensual, powerful, able to recreate myself with every new day." But no hard feelings toward those who choose to hide their condition under wigs or scarves; it's just a matter, *Mamm* tells us, of "different aesthetics." Some go for pink ribbons; others will prefer the Ralph Lauren Pink Pony breast-cancer motif. But everyone agrees that breast cancer is a chance for creative self-transformation — a makeover opportunity, in fact.

Now, cheerfulness, up to and including delusion and false hope, has a recognized place in medicine. There is plenty of evidence that depressed and socially isolated people are more prone to succumb to diseases, cancer included, and a diagnosis of cancer is probably capable of precipitating serious depression all by itself. To be told by authoritative figures that you have a deadly disease, for which no real cure exists, is to enter a liminal state fraught with perils that go well beyond the disease itself. Consider the phenomenon of "voodoo death" — described by ethnographers among, for example, Australian aborigines — in which a person who has been condemned by a suitably potent curse obligingly shuts down and dies within a day or two. Cancer diagnoses could, and in some cases probably do, have the same kind of fatally dispiriting effect. So, it could be argued, the collectively pumped-up optimism of breast-cancer culture may be just what the doctor ordered. Shop for the Cure, dress in pink-ribbon regalia, organize a run or hike — whatever gets you through the night.

But in the seamless world of breast-cancer culture, where one Web site links to another — from personal narratives and grassroots endeavors to the glitzy level of corporate sponsors and celebrity spokespeople — cheerfulness is more or less mandatory, dissent a kind of treason. Within this tightly knit world, attitudes are subtly adjusted, doubters gently brought back to the fold. In *The First Year of the Rest of Your Life,* for example, each personal narrative is followed by a study question or tip designed to counter the slightest hint of negativity — and they are very slight hints indeed, since the collection includes no harridans, whiners, or feminist militants:

> Have you given yourself permission to acknowledge you have some anxiety or "blues" and to ask for help for your emotional well-being?
>
> Is there an area in your life of unresolved internal conflict? Is there an area where you think you might want to do some "healthy mourning"?
>
> Try keeping a list of the things you find "good about today."

As an experiment, I post a statement on the Komen.org message board, under the subject line "angry," briefly listing my own heartfelt complaints about debilitating treatments, recalcitrant insurance companies, environmental carcinogens, and, most daringly, "sappy pink ribbons." I receive a few words of encouragement in my fight with the insurance company, which has taken the position that my biopsy was a kind of optional indulgence, but mostly a chorus of rebukes. "Suzy" writes to say, "I really dislike saying you have a bad attitude towards all of this, but you do, and it's not going to help you in the least." "Mary" is a bit more tolerant, writing, "Barb, at this time in your life, it's so important to put all your energies toward a peaceful, if not happy, existence. Cancer is a rotten thing to have happen and there are no answers for any of us as to why. But to live your life, whether you have one more year or 51, in anger and bitterness is such a waste . . . I hope you can find some peace. You deserve it. We all do. God bless you and keep you in His loving care. Your sister, Mary."

"Kitty," however, thinks I've gone around the bend: "You need to run, not walk, to some counseling. . . . Please, get yourself some help and I ask everyone on this site to pray for you so you can enjoy life to the fullest."

I do get some reinforcement from "Gerri," who has been through all the treatments and now finds herself in terminal condition: "I am also angry. All the money that is raised, all the smiling faces of survivors who make it sound like it is o.k. to have breast cancer. IT IS NOT O.K.!" But Gerri's message, like the others on the message board, is posted under the mocking heading "What does it mean to be a breast cancer survivor?"

Culture is too weak a word to describe all this. What has grown up around breast cancer in just the last fifteen years more nearly resembles a cult — or, given that it numbers more than 2 million women, their families, and friends — perhaps we should say a full-fledged religion. The products — teddy bears, pink-ribbon brooches, and so forth — serve as amulets and talismans, comforting the sufferer and providing visible evidence of faith. The personal narratives serve as testimonials and follow the same general arc as the confessional autobiographies required of seventeenth-century Puritans: first there is a crisis, often involving a sudden apprehension of mortality (the diagnosis or, in the old Puritan case, a stern word from on high); then comes a prolonged ordeal (the treatment or, in the religious case, internal struggle with the Devil); and finally, the blessed certainty of salvation, or its breast-cancer equivalent, survivorhood. And like most recognized religions, breast cancer has its great epideictic events, its pilgrimages and mass gatherings where the faithful convene and draw strength from their numbers. These are the annual races for a cure, attracting a total of about a million people at more than eighty sites — seventy thousand of them at the largest event, in Washington, D.C., which in recent years has been attended by Dan and Marilyn

Quayle and Al and Tipper Gore. Everything comes together at the races: celebrities and corporate sponsors are showcased; products are hawked; talents, like those of the "Swinging, Singing Survivors" from Syracuse, New York, are displayed. It is at the races, too, that the elect confirm their special status. As one participant wrote in the *Washington Post:*

> I have taken my "battle scarred" breasts to the Mall, donned the pink shirt, visor, pink shoelaces, etc. and walked proudly among my fellow veterans of the breast cancer war. In 1995, at the age of 44, I was diagnosed and treated for Stage II breast cancer. The experience continues to redefine my life.

Feminist breast-cancer activists, who in the early nineties were organizing their own mass outdoor events — demonstrations, not races — to demand increased federal funding for research, tend to keep their distance from these huge, corporate-sponsored, pink gatherings. Ellen Leopold, for example — a member of the Women's Community Cancer Project in Cambridge and author of *A Darker Ribbon: Breast Cancer, Women, and Their Doctors in the Twentieth Century* — has criticized the races as an inefficient way of raising money. She points out that the Avon Breast Cancer Crusade, which sponsors three-day, sixty-mile walks, spends more than a third of the money raised on overhead and advertising, and Komen may similarly fritter away up to 25 percent of its gross. At least one corporate-charity insider agrees. "It would be much easier and more productive," says Rob Wilson, an organizer of charitable races for corporate clients, "if people, instead of running or riding, would write out a check to the charity."

To true believers, such criticisms miss the point, which is always, ultimately, "awareness." Whatever you do to publicize the disease — wear a pink ribbon, buy a teddy, attend a race — reminds other women to come forward for their mammograms. Hence, too, they would argue, the cult of the "survivor": if women neglect their annual screenings, it must be because they are afraid that a diagnosis amounts to a death sentence. Beaming survivors, proudly displaying their athletic prowess, are the best possible advertisement for routine screening mammograms, early detection, and the ensuing round of treatments. Yes, miscellaneous businesses — from tiny distributors of breast-cancer wind chimes and note cards to major corporations seeking a woman-friendly image — benefit in the process, not to mention the breast-cancer industry itself, the estimated $12 to $16 billion-a-year business in surgery, "breast health centers," chemotherapy "infusion suites," radiation treatment centers, mammograms, and drugs ranging from antiemetics (to help you survive the nausea of chemotherapy) to tamoxifen (the hormonal treatment for women with estrogen-sensitive tumors). But what's to complain about? Seen through pink-tinted lenses, the entire breast-cancer enterprise — from grassroots support groups and Web sites to the corporate providers of therapies and sponsors of races — looks like a beautiful example of synergy at work: cult activities, paraphernalia, and testimonies encourage women to undergo the diagnostic procedures, and since a fraction of these diagnoses will be positive, this means more members for the cult as well as more customers for the corporations, both those that provide medical products and services and those that offer charitable sponsorships.

But this view of a life-giving synergy is only as sound as the science of current detection and treatment modalities, and, tragically, that science is fraught with doubt, dissension, and what sometimes looks very much like denial. Routine screening mammograms, for example, are the major goal of "awareness," as when Rosie O'Donnell exhorts us to go out and "get squished." But not all breast-cancer experts are as enthusiastic. At best the evidence for the salutary effects of routine mammograms — as opposed to breast self-examination — is equivocal, with many respectable large-scale studies showing a vanishingly small impact on overall breast-cancer mortality. For one thing, there are an estimated two to four false positives for every cancer detected, leading thousands of healthy women to go through unnecessary biopsies and anxiety. And even if mammograms were 100 percent accurate, the admirable goal of "early" detection is more elusive than the current breast-cancer dogma admits. A small tumor, detectable only by mammogram, is not necessarily young and innocuous; if it has not spread to the lymph nodes, which is the only form of spreading detected in the common surgical procedure of lymph-node dissection, it may have already moved on to colonize other organs via the bloodstream. David Plotkin, director of the Memorial Cancer Research Foundation of Southern California, concludes that the benefits of routine mammography "are not well established; if they do exist, they are not as great as many women hope." Alan Spievack, a surgeon recently retired from the Harvard Medical School, goes further, concluding from his analysis of dozens of studies that routine screening mammography is, in the words of famous British surgeon Dr. Michael Baum, "one of the greatest deceptions perpetrated on the women of the Western world."

Even if foolproof methods for early detection existed, they would, at the present time, serve only as portals to treatments offering dubious protection and considerable collateral damage. (Some improved prognostic tools, involving measuring a tumor's growth rate and the extent to which it is supplied with blood vessels, are being developed but are not yet in use.) Some women diagnosed with breast cancer will live long enough to die of something else, and some of these lucky ones will indeed owe their longevity to a combination of surgery, chemotherapy, radiation, and/or anti-estrogen drugs such as tamoxifen. Others, though, would have lived untreated or with surgical excision alone, either because their cancers were slow-growing or because their bodies' own defenses were successful. Still others will die of the disease no matter what heroic, cell-destroying therapies are applied. The trouble is, we do not have the means to distinguish between these three groups. So for many of the thousands of women who are diagnosed each year, Plotkin notes, "the sole effect of early detection has been to stretch out the time in which the woman bears the knowledge of her condition." These women do not live longer than they might have without any medical intervention, but more of the time they do live is overshadowed with the threat of death and wasted in debilitating treatments.

To the extent that current methods of detection and treatment fail or fall short, America's breast-cancer cult can be judged as an outbreak of mass delusion, celebrating survivorhood by downplaying mortality and promoting obedience to medical protocols known to have limited efficacy. And although we may imagine ourselves to be well past the era of patriarchal medicine, obedience is the message behind the infantilizing theme in breast-cancer culture, as represented by the teddy bears, the crayons, and the prevailing pinkness. You are encouraged to regress to a little-girl state, to suspend critical judgment, and to accept whatever measures the doctors, as parent surrogates, choose to impose.

Worse, by ignoring or underemphasizing the vexing issue of environmental causes, the breast-cancer cult turns women into dupes of what could be called the Cancer Industrial Complex: the multinational corporate enterprise that with the one hand doles out carcinogens and disease and, with the other, offers expensive, semitoxic pharmaceutical treatments. Breast Cancer Awareness Month, for example, is sponsored by AstraZeneca (the manufacturer of tamoxifen), which, until a corporate reorganization in 2000, was a leading producer of pesticides, including acetochlor, classified by the EPA as a "probable human carcinogen." This particularly nasty conjuncture of interests led the environmentally oriented Cancer Prevention Coalition (CPC) to condemn Breast Cancer Awareness Month as "a public relations invention by a major polluter which puts women in the position of being unwitting allies of the very people who make them sick." Although AstraZeneca no longer manufactures pesticides, CPC has continued to criticize the breast-cancer crusade — and the American Cancer Society — for its unquestioning faith in screening mammograms and careful avoidance of environmental issues. In a June 12, 2001, press release, CPC chairman Samuel S. Epstein, M.D., and the well-known physician activist Quentin Young castigated the American Cancer Society for its "longstanding track record of indifference and even hostility to cancer prevention. . . . Recent examples include issuing a joint statement with the Chlorine Institute justifying the continued global use of persistent organochlorine pesticides, and also supporting the industry in trivializing dietary pesticide residues as avoidable risks of childhood cancer. ACS policies are further exemplified by allocating under 0.1 percent of its $700 million annual budget to environmental and occupational causes of cancer."

In the harshest judgment, the breast-cancer cult serves as an accomplice in global poisoning — normalizing cancer, prettying it up, even presenting it, perversely, as a positive and enviable experience.

When, my three months of chemotherapy completed, the oncology nurse calls to congratulate me on my "excellent blood work results," I modestly demur. I didn't do anything, I tell her, anything but endure — marking the days off on the calendar, living on Protein Revolution canned vanilla health shakes, escaping into novels and work. Courtesy restrains me from mentioning the fact that the tumor markers she's tested for have little prognostic value, that there's no way to know how many rebel cells survived chemotherapy and may be carving out new colonies right now. She insists I should be proud; I'm a survivor now and entitled to recognition at the Relay for Life being held that very evening in town.

So I show up at the middle-school track where the relay's going on just in time for the Survivors' March: about one hundred people, including a few men, since the funds raised will go to cancer research in general, are marching around the track eight to twelve abreast while a loudspeaker announces their names and survival times and a thin line of observers, mostly people staffing the raffle and food booths, applauds. It could be almost any kind of festivity, except for the distinctive stacks of cellophane-wrapped pink Hope Bears for sale in some of the booths. I cannot help but like the funky small-town *Gemütlichkeit* of the event, especially when the audio system strikes up that universal anthem of solidarity, "We Are Family," and a few people of various ages start twisting to the music on the jerry-built stage. But the money raised is going far away, to the American Cancer Society, which will not be asking us for our advice on how to spend it.

I approach a woman I know from other settings, one of our local intellectuals, as it happens, decked out here in a pink-and-yellow survivor T-shirt and with an American Cancer Society "survivor medal" suspended on a purple ribbon around her neck. "When do you date your survivorship from?" I ask her, since the announced time, five and a half years, seems longer than I recall. "From diagnosis or the completion of your treatments?" The question seems to annoy or confuse her, so I do not press on to what I really want to ask: at what point, in a downwardly sloping breast-cancer career, does one put aside one's survivor regalia and admit to being in fact a die-er? For the dead are with us even here, though in much diminished form. A series of paper bags, each about the right size for a junior burger and fries, lines the track. On them are the names of the dead, and inside each is a candle that will be lit later, after dark, when the actual relay race begins.

My friend introduces me to a knot of other women in survivor gear, breast-cancer victims all, I learn, though of course I would not use the V-word here. "Does anyone else have trouble with the term *survivor*?" I ask, and, surprisingly, two or three speak up. It could be "unlucky," one tells me; it "tempts fate," says another, shuddering slightly. After all, the cancer can recur at any time, either in the breast or in some more strategic site. No one brings up my own objection to the term, though: that the mindless triumphalism of "survivorhood" denigrates the dead and the dying. Did we who live "fight" harder than those who've died? Can we claim to be "braver," better, people than the dead? And why is there no room in this cult for some gracious acceptance of death, when the time comes, which it surely will, through cancer or some other misfortune?

No, this is not my sisterhood. For me at least, breast cancer will never be a source of identity or pride. As my dying correspondent Gerri wrote: "IT IS NOT O.K.!" What it is, along with cancer generally or any slow and painful way of dying, is an abomination, and, to the extent that it's man-made, also a crime. This is the one great truth that I bring out of the breast-cancer experience, which did not, I can now report, make me prettier or stronger, more feminine or spiritual — only more deeply angry. What sustained me through the "treatments" is a purifying rage, a resolve, framed in the sleepless nights of chemotherapy, to see the last polluter, along with, say, the last smug health-insurance operative, strangled with the last pink ribbon. Cancer or no cancer, I will not live that long, of course. But I know this much right now for sure: I will not go into that last good night with a teddy bear tucked under my arm.

H. BRUCE FRANKLIN

The Most Important Fish in the Sea

FROM *Discover*

FIRST YOU SEE THE BIRDS — gulls, terns, cormorants, and ospreys wheeling overhead, then swooping down into a wide expanse of water dimpled as though by large raindrops. Silvery flashes and splashes erupt from thousands of small herring-like fish called menhaden. More birds arrive, and the air rings with shrill cries. The birds alert nearby anglers that a massive school of menhaden is under attack by bluefish.

The razor-toothed blues tear at the menhaden like piranhas in a killing frenzy, gorging themselves, some killing even when they are too full to eat, some vomiting so they can kill and eat again. Beneath the blues, weakfish begin to circle, snaring the detritus of the carnage. Farther below, giant striped bass gobble chunks that get by the weakfish. From time to time a bass muscles its way up through the blues to take in whole menhaden. On the sea floor, scavenging crabs feast on leftovers.

The school of menhaden survives and swims on, its losses dwarfed in plenitude. But a greater danger than bluefish lurks nearby. The birds have attracted a spotter-plane pilot who works for Omega Protein, a $100 million fishing corporation devoted entirely to catching menhaden. As the pilot approaches, he sees the school as a neatly defined silver-purple mass the size of a football field and perhaps 100 feet deep. He radios to a nearby 170-foot-long factory ship, whose crew maneuvers close enough to launch two 40-foot-long boats. The pilot directs the boats' crews as they deploy a purse seine, a gigantic net. Before long, the two boats have trapped the entire school. As the fish strike the net, they thrash frantically, making a wall of white froth that marks the net's circumference. The factory ship pulls alongside, pumps the fish into its refrigerated hold, and heads off to unload them at an Omega plant in Virginia.

Not one of these fish is destined for a supermarket, canning factory, or restaurant. Menhaden are oily and foul and packed with tiny bones. No one eats them. Yet they are the most important fish caught along the Atlantic and Gulf coasts, exceeding the tonnage of all other species combined. These kibble of the sea fetch only about ten cents a pound at the dock, but they can be ground up, dried, and formed into another kind of kibble for land animals, a high-protein feed for chickens, pigs, and cattle. Pop some barbecued wings into your mouth, and at least part of what you're eating was once menhaden.

Humans eat menhaden in other forms, too. Menhaden are a key dietary component for a wide variety of fish, including bass, mackerel, cod, bonito, swordfish, bluefish, and tuna. The nineteenth-century ichthyologist G. Brown Goode exaggerated only slightly when declaring that people who dine on Atlantic saltwater fish are eating "nothing but menhaden."

And that is one problem with the intensive fishing of menhaden, which has escalated in recent decades. This vital biolink in a food chain that extends from tiny plankton to the dinner tables of many Americans appears to be threatened. The population of menhaden has been so depleted in estuaries and bays up and down the Eastern Seaboard that even marine biologists who look kindly on commercial fishing are alarmed. "Menhaden are an incredibly important link for the entire Atlantic coast," says Jim Uphoff, the stock assessment coordinator for the Fisheries Service of the Maryland Department of Natural Resources. "And you have a crashing menhaden population with the potential to cause a major ecosystem problem." Menhaden have an even more important role that extends beyond the food chain: they are filter feeders that consume phytoplankton, thus controlling the growth of algae in coastal waters. As the population of menhaden declines, algal blooms have proliferated, transforming some inshore waters into dead zones.

To grasp how ubiquitous menhaden once were, you can read the journals of explorer John Smith. In 1607, he sailed across the

Chesapeake Bay through a mass of menhaden he described as "lying so thick with their heads above the water, as for want of nets (our barge driving amongst them) we attempted to catch them with a frying pan." Colossal schools of menhaden, often more than a mile in diameter, were once common along the entire Atlantic and Gulf coasts of the United States. Since World War II, however, fishermen using spotter planes and purse seines appear to have dramatically decreased both the population and the range of menhaden.

Bryan Taplin, an environmental scientist in the Atlantic Ecology Division of the Environmental Protection Agency (EPA), has witnessed the destruction of all the large schools of menhaden by purse seiners in Rhode Island's Narragansett Bay. During the last two decades he has also studied changes in the diet of striped bass in the bay by analyzing the carbon isotope signature of their scales. What he has discovered is a steady shift away from fat-rich menhaden to invertebrates that provide considerably lower nutritional value. That has been accompanied by a loss of muscle and a decrease in the weight-to-length ratio of striped bass. The bass that remain in Narragansett Bay, says Taplin, are "long skinny stripers" that have been forced to shift their diet because "the menhaden population has crashed to an all-time low."

"You have to scratch your head and wonder — since we set quotas for bluefin and tuna — why we don't set quotas for this crucial part of the oceanic food chain," says Taplin. "Not to regulate a fishery that's so important is to ask for trouble. I wonder whether we are about to see something go wrong unlike anything we have ever seen."

Signs of what could go wrong are already obvious in the Chesapeake Bay, the tidal estuary that once produced more seafood per acre than any body of water on earth. "There's nothing in Chesapeake Bay that can take the place of menhaden," says Uphoff of the Maryland Fisheries Service. "Menhaden are king." Jim Price is a fifth-generation Chesapeake Bay fisherman. For ten years he captained a charter boat specializing in light-tackle fishing for striped bass, also called rockfish by bay anglers. One day in the fall of 1997, Price caught a rockfish so diseased he still becomes upset when he talks about it. "I'd never seen anything like that in my entire life," he says, wringing his powerful, deeply tanned hands. "It was covered with red sores. It was so sickening it really took something out of me."

Price deposited several sick rockfish at the Cooperative Oxford Laboratory in nearby Oxford, Maryland, and then began his own independent study. When he cut some open, he was shocked. "I've been looking in the stomachs of rockfish for forty years," he says, "but I couldn't believe what I saw — nothing, absolutely nothing. Not only was there no food, but there was no fat. Everything was shrunk up and small."

An Oxford lab pathologist speculated that the fish might have been "decoupled from their source of food," but Price was incredulous. "I thought to myself, with all the food here in the Chesapeake, that's a stupid idea. Then I got to thinking. In years past, at that time of year I would find their stomachs full of menhaden, sometimes a half-dozen whole fish."

Price hypothesized that malnutrition, caused by the decline in the menhaden population, made the rockfish vulnerable to disease. Since then, his hypothesis has been confirmed by research. Half the rockfish in the Chesapeake are diseased, with either bacterial infections or lesions associated with *Pfiesteria,* a toxic form of phytoplankton known as the cell from hell. But that is only one symptom of the depletion of menhaden.

Dense schools of menhaden swimming with their mouths open slurp up enormous quantities of plankton and detritus like gargantuan vacuum cleaners. In the Chesapeake and other coastal waterways, the filtering clarifies water by purging suspended particles that cause turbidity, allowing sunlight to penetrate to greater depths. That encourages the growth of plants that release dissolved oxygen as they photosynthesize. The plants also harbor fish and shellfish.

Far more important, the menhaden's filter feeding limits the spread of devastating algal blooms. Runoff from many sources — farms, detergent-laden wastewater, overfertilized golf courses, and suburban lawns — floods nitrogen and phosphorus into coastal waters. Nitrogen and phosphorus in turn stimulate the growth of algal blooms that block sunlight and kill fish. The blooms eventually sink in thick carpets to the sea bottom, where they suck dissolved oxygen from the water and leave dead zones. Menhaden, by consuming nutrient-rich phytoplankton and then either swimming out to sea in seasonal migrations or being consumed by fish, birds, and marine mammals, remove a significant percentage of the excess nitrogen and phosphorus that cause algal overgrowth.

Nature had developed a marvelous method for keeping bays and estuaries clear, clean, balanced, and healthy: oysters, the other great filter feeders, removed plankton in lower water layers, and menhaden removed it from upper layers. As oysters have been driven to near extinction along parts of the Atlantic coast, menhaden have become increasingly important as filters.

Marine biologist Sara Gottlieb says: "Think of menhaden as the liver of a bay. Just as your body needs its liver to filter out toxins, ecosystems also need those natural filters." Overfishing of menhaden is "just like removing your liver," she says, and "you can't survive without a liver."

During the late nineteenth century, several dozen sailing vessels and a handful of steamships hunted menhaden in Gardiners Bay, near the eastern tip of Long Island, New York. The abundance of menhaden then appealed to another set of hunters: ospreys that nested in an immense rookery on Gardiners Island. As late as the mid-1940s, there were still three hundred active osprey nests on the small island. But the ospreys fell victim to the DDT that was sprayed on the wetlands. Eventually, the number of active nests plummeted to twenty-six. After DDT was banned, biologist Paul Spitzer observed a gradual resurgence of the osprey. However, in recent years he has watched the number of ospreys on Gardiners Island dwindle again. From 1995 to 2001, he says, "there has been an absolute steep decline from seventy-one active nests to thirty-six."

Although no longer weakened by toxins, ospreys now have little to eat. "Migratory menhaden schools formerly arrived in May, in time to feed nestlings," Spitzer says. In recent years, menhaden have disappeared, and the survival rate of osprey chicks has fallen to one chick for every two nests, a rate comparable to the worst years of DDT use. "The collapse of the menhaden means the endgame for Gardiners Island ospreys," he says. Spitzer sees the same pattern of decline in other famous osprey colonies, including those at Plum Island, Massachusetts; Cape Henlopen, Delaware; Smith Point, New York; and Sandy Hook and Cape May in New Jersey.

The menhaden crash may also contribute to the decline of the loons that make an autumn migration stopover in the Chesapeake each year. Spitzer keeps statistical counts of flocks passing through a roughly sixty-square-mile prime habitat on the Chesapeake's Choptank River, near where Jim Price found diseased striped bass. Between 1989 and 1999, Spitzer's loon count dropped steadily from 750 to 1,000 per three-hour observation period to 75 to 200. The typical flock fell from 100 to 500 birds to between 15 and 40. Menhaden are "the absolute keystone species for the health of the entire Atlantic ecosystem," says Spitzer.

Hall Watters, now seventy-six and retired, looks back ruefully on the role he and other spotter pilots played in the demise of the menhaden. "We are what destroyed the fishery, because the menhaden had no place to hide," he says. "If you took the airplanes away from the fleet, the fish would come back."

Watters was the first menhaden spotter pilot, hired in 1946 by Brunswick Navigation of Southport, North Carolina. He had been a fighter pilot during World War II and says he was "the only pilot around who knew what menhaden looked like." Brunswick had just converted three oceangoing minesweepers and two submarine chasers to menhaden fishing ships and was eager to extend the range and efficiency of its operations. Menhaden usually spawn far out at sea, and the larvae must be carried by currents to the inshore waterways where they mature. Guided by Watters, Brunswick's rugged vessels soon began to net schools as far out as fifty miles, some with so many egg-filled females, he says, that the nets "would be all slimy from the roe."

Watters remembers that in the early postwar years, menhaden filled the seas. In 1947, he spotted one school about fifteen miles off Cape Hatteras so large that from an altitude of 10,000 feet, it looked like an island. Although a hundred boats circled the school, many fish escaped. "Back then we only fished the big schools. We used to stop when the schools broke up into small pods." But things had changed dramatically by the time he quit in 1980: "We caught everything we saw. The companies wanted to catch everything but the wiggle."

The exact size of the Atlantic menhaden population in 2001 is impossible to measure, but industry statistics show a dramatic decline in catches over the years since 1946. The average annual tonnage from 1996 to 1999 was only 40 percent of the average annual tonnage caught between 1955 and 1961. Last year the catch was the second lowest in sixty years. Moreover, these numbers may not reflect the full scope of the decline because the catch is not necessarily proportional to the population. "The stock gets smaller but still tends to school," says Jim Uphoff of the Maryland Fisheries Service. "The fishery gets more efficient at finding the schools. Thus they take a larger fraction of the population as the stock is going down."

The large oceanic schools of menhaden are often too scarce to chase profitably, so the fishing industry has moved into estuaries and bays, particularly the Chesapeake. Maryland has banned purse seining in its portion of the Chesapeake. Virginia has not. Omega Protein, headquartered in Houston and the largest U.S. menhaden fishing firm, has almost unlimited access to state waters, including the mouth and southern half of the Chesapeake. By 1999, 60 percent of the entire Atlantic menhaden catch came from the Virginia waters of the Chesapeake.

These days Omega Protein enjoys a near monopoly fishing for menhaden. As the fish population declined and operational costs increased, many companies went bankrupt or were bought out by bigger, more industrialized corporations. Omega Protein's parent was Zapata, a Houston-based corporation cofounded by former president George Bush in 1953. Omega Protein went independent in 1998, after completing the consolidation of the menhaden industry by taking over its large Atlantic competitor, American Protein of Virginia, and its Gulf competitor, Gulf Protein of Louisiana.

Omega Protein mothballed thirteen of its fifty-three ships last year and grounded twelve of its forty-five spotter planes as the menhaden continued to disappear. Fewer than a dozen of the company's ships fish out of Virginia, but thirty ships fish the Gulf of Mexico.

The Gulf seems to be headed for the same problems that are obvious in the Chesapeake, but on a larger scale. Fed by chemical runoff, algal blooms have spread, causing ever-enlarging, oxygen-depleted dead zones. And jellyfish are proliferating, both a native species and a gigantic Pacific species. Researchers believe the swollen jellyfish population could have a devastating effect on Gulf fishing because they attack the eggs and larvae of many species.

Monty Graham, senior marine scientist at the Dauphin Island Sea Lab in Alabama, says overfishing, "including aggressive menhaden fishing," seems to have allowed the jellyfish — "an opportunistic planktivore" — to fill the ecological void. He says the proliferation of both species of jellyfish indicates "something gone wrong with the ecology."

Barney White, corporate vice president of Omega Protein and chairman of the National Fish Meal and Oil Association, the industry's trade association, categorically denies that menhaden are being overfished or that there is any ecological problem whatsoever caused by their decline. He says the controversy "is largely without basis" and is based on "lies" disseminated by recreational fishermen in general and Jim Price in particular. "It becomes an issue of politics rather than science — that people have a problem with commercial fishing in general," White says. "We have big boats closer to shore, so we're easy to see, and that makes us a convenient political target."

White attributes the absence of adult fish in New England and eastern Long Island waters to cyclic factors. "Well-meaning people who don't know marine biology have been mistaking short-term occurrences for long-term trends," he says. "In fact, the reports I have show that more fish seem to be moving into the area." Moreover, White says, "the total biomass is sufficient to sustain the industry."

Watters disagrees. More than a half century after he first took to the air as a spotter pilot, he fumes that "the industry destroyed their own fishery, and they're still at it." What galls him the most is that an increasing proportion of the catch consists of "zeros" — menhaden less than a year old. He advocates banning menhaden fishing close to shore, especially in estuaries, where the young menhaden mature. He also argues that if Omega Protein "enlarged the mesh size, they wouldn't be wiping out the zero class."

White acknowledges the industry is facing a problem of "recruitment" — menhaden are not living through their first year. But he insists that the one-and-three-quarter-inch mesh now used allows the very smallest juveniles to slip through. The real problem, he says, is "an overpopulation of striped bass. We think the striped bass are eating all the juveniles."

Omega Protein's financial reports indicate that the fortunes of the company rise and fall with "the supply and demand for competing products, particularly soybean meal for its fish meal products and vegetable oils and fats for its fish oil products." The fishing industry's journal, *National Fisherman,* says: "On the industrial side of the fishery, where menhaden is processed into feed for poultry and pigs, the demand for fish is depressed by a surplus of soy, which serves the same purpose." In other words, all the ground-up menhaden could be replaced by ground-up soybeans.

Since market forces are unlikely to curtail the menhaden fishery, governments may have to take action. Price thinks the fishing season for menhaden should be closed each December 1, "because after that is when the age zeros migrate down the coast." No matter what is done, most researchers agree the menhaden must be viewed not as a specific problem about a single species of disappearing fish but as a much larger ecological threat.

Bill Matuszeski, former executive director of the National Marine Fisheries Service and former director of the EPA's Chesapeake Bay program, says: "We need to start managing menhaden for their role in the overall ecological system. If this problem isn't taken care of, the EPA will have to get into the decision making." Matuszeski believes estuaries like the Chesapeake Bay should be put off limits to menhaden fishing immediately. "That would be inconvenient for the industry, but it would be inconvenient for the species to be extinct."

MALCOLM GLADWELL

Examined Life

FROM *The New Yorker*

ONCE, IN FOURTH GRADE, Stanley Kaplan got a B-plus on his report card and was so stunned that he wandered aimlessly around the neighborhood, ashamed to show his mother. This was in Brooklyn, on Avenue K in Flatbush, between the wars. Kaplan's father, Julius, was from Slutsk, in Belorussia, and ran a plumbing and heating business. His mother, Ericka, ninety pounds and four feet eight, was the granddaughter of the chief rabbi of the synagogue of Prague, and Stanley loved to sit next to her on the front porch, immersed in his schoolbooks while his friends were off playing stickball. Stanley Kaplan had Mrs. Holman for fifth grade, and when she quizzed the class on math equations, he would shout out the answers. If other students were having problems, Stanley would take out pencil and paper and pull them aside. He would offer them a dime sometimes if they would just sit and listen. In high school, he would take over algebra class, and the other kids, passing him in the hall, would call him Teach. One classmate, Aimee Rubin, was having so much trouble with math that she was in danger of being dropped from the National Honor Society. Kaplan offered to help her, and she scored a ninety-five on her next exam. He tutored a troubled eleven-year-old named Bob Linker, and Bob Linker ended up a successful businessman. In Kaplan's sophomore year at City College, he got a C in biology and was so certain that there had been a mistake that he marched in to see the professor and proved that his true grade, an A, had accidentally been switched with that of another, not quite so studious Stanley Kaplan. Thereafter, he became Stanley H. Kaplan, and when people asked him what the "H" stood for, he would say "Higher scores!" or, with a sly wink, "Preparation!" He graduated Phi Beta Kappa and hung a shingle outside his parent's house on Avenue K — "Stanley H. Kaplan Educational Center" — and started tutoring kids in the basement. In 1946, a high school junior named Elizabeth, from Coney Island, came to him for help on an exam he was unfamiliar with. It was called the Scholastic Aptitude Test, and from that moment forward the business of getting into college in America was never quite the same.

The SAT, at that point, was just beginning to go into widespread use. Unlike existing academic exams, it was intended to measure innate ability — not what a student had learned but what a student was capable of learning — and it stated clearly in the instructions that "cramming or last-minute reviewing" was pointless. Kaplan was puzzled. In Flatbush you always studied for tests. He gave Elizabeth pages of math problems and reading-comprehension drills. He grilled her over and over, doing what the SAT said should not be done. And what happened? On test day, she found the SAT "a piece of cake," and promptly told all her friends, and her friends told their friends, and soon word of Stanley H. Kaplan had spread throughout Brooklyn.

A few years later, Kaplan married Rita Gwirtzman, who had grown up a mile away, and in 1951 they moved to a two-story brick-and-stucco house on Bedford Avenue, a block from his alma mater, James Madison High School. He renovated his basement, dividing it into classrooms. When the basement got too crowded, he rented a podiatrist's office near King's Highway, at the Brighton Beach subway stop. In the 1970s, he went national, setting up educational programs throughout the country, creating an SAT-preparation industry that soon became crowded with tutoring companies and study manuals. Kaplan has now written a memoir, *Test Pilot,* which has as its subtitle *How I Broke Testing Barriers for Millions of Students and Caused a Sonic Boom in the Business of Education.* That actually understates his importance. Stanley Kaplan changed the rules of the game.

The SAT is now seventy-five years old, and it is in trouble. Earlier this year, the University of California — the nation's largest public university system — stunned the educational world by proposing a move toward a "holistic" admissions system, which would mean abandoning its heavy reliance on standardized-test scores. The school backed up its proposal with a devastating statistical analysis, arguing that the SAT is virtually useless as a tool for making admissions decisions.

The report focused on what is called predictive validity, a statistical measure of how well a high school student's performance in any given test or program predicts his or her performance as a college freshman. If you wanted to, for instance, you could calculate the predictive validity of prowess at Scrabble, or the number of books a student reads in his senior year, or, more obviously, high school grades. What the Educational Testing Service (which creates the SAT) and the College Board (which oversees it) have always argued is that most performance measures are so subjective and unreliable that only by adding aptitude-test scores into the admissions equation can a college be sure it is picking the right students.

This is what the UC study disputed. It compared the predictive validity of three numbers: a student's high school GPA, his or her score on the SAT (or, as it is formally known, the SAT I), and his or her score on what is known as the SAT II, which is a so-called achievement test, aimed at gauging mastery of specific areas of the high school curriculum. Drawing on the transcripts of 78,000 University of California freshmen from 1996 through 1999, the report found that overall, the most useful statistic in predicting freshman grades was the SAT II, which explained 16 percent of the "variance" (which is another measure of predictive validity). The second most useful was high school GPA, at 15.4 percent. The SAT was the least useful, at 13.3 percent. Combining high school GPA and the SAT II explained 22.2 percent of the variance in freshman grades. Adding in SAT I scores increased that number by only 0.1 percent. Nor was the SAT better at what one would have thought was its strong suit: identifying high-potential students from bad schools. In fact, the study found that achievement tests were ten times more useful than the SAT in predicting the success of students from similar backgrounds. "Achievement tests are fairer to students because they measure accomplishment rather than promise," Richard Atkinson, the president of the University of California, told a conference on college admissions last month. "They can be used to improve performance; they are less vulnerable to charges of cultural or socioeconomic bias; and they are more appropriate for schools because they set clear curricular guidelines and clarify what is important for students to learn. Most important, they tell students that a college education is within the reach of anyone with the talent and determination to succeed."

This argument has been made before, of course. The SAT has been under attack, for one reason or another, since its inception. But what is happening now is different. The University of California is one of the largest single customers of the SAT. It was the UC system's decision, in 1968, to adopt the SAT that affirmed the test's national prominence in the first place. If UC defects from the SAT, it is not hard to imagine it being followed by a stampede of other colleges. Seventy-five years ago, the SAT was instituted because we were more interested, as a society, in what a student was capable of learning than in what he had already learned. Now, apparently, we have changed our minds, and few people bear more responsibility for that shift than Stanley H. Kaplan.

From the moment he set up shop on Avenue K, Stanley Kaplan was a pariah in the educational world. Once, in 1956, he went to a meeting for parents and teachers at a local high school to discuss the upcoming SAT, and one of the teachers leading the meeting pointed his finger at Kaplan and shouted, "I refuse to continue until *that man* leaves the room." When Kaplan claimed that his students routinely improved their scores by a hundred points or more, he was denounced by the testing establishment as a "quack" and "the cram king" and a "snake oil salesman." At the Educational Testing Service, "it was a cherished assumption that the SAT was uncoachable," Nicholas Lemann writes in his history of the SAT, *The Big Test:*

> The whole idea of psychometrics was that mental tests are a measurement of a psychical property of the brain, analogous to taking a blood sample. By definition, the test-taker could not affect the result. More particularly, ETS's main point of pride about the SAT was its extremely high test-retest reliability, one of the best that any standardized test had ever achieved. . . . So confident of the SAT's reliability was ETS that the basic technique it developed for catching cheaters was simply to compare first and second scores, and to mount an investigation in the case of any very large increase. ETS was sure that substantially increasing one's score could be accomplished only by nefarious means.

But Kaplan wasn't cheating. His great contribution was to prove that the SAT was eminently coachable — that whatever it was that the test was measuring was less like a blood sample than like a heart rate, a vital sign that could be altered through the right exercises. In those days, for instance, the test was a secret. Students walking in to take the SAT were often in a state of terrified ignorance about what to expect. (It wasn't until the early eighties that the ETS was forced to release copies of old test questions to the public.) So Kaplan would have "Thank Goodness It's Over" pizza parties after each SAT. As his students talked about the questions they had faced, he and his staff would listen and take notes, trying to get a sense of how better to structure their coaching. "Every night I stayed up past midnight writing new questions and study materials," he writes. "I spent hours trying to understand the design of the test, trying to think like the test makers, anticipating the types of questions my students would face." His notes were typed up the next day, cranked out on a Gestetner machine, hung to dry in the office, then snatched off the line and given to waiting students. If students knew what the SAT was like, he reasoned, they would be more confident. They could skip the instructions and save time. They could learn how to pace themselves. They would guess more intelligently. (For a question with five choices, a right answer is worth one point but a wrong answer results in minus one-quarter of a point — which is why students were always warned that guessing was penalized. In reality, of course, if a student can eliminate even one obviously wrong possibility from the list of choices, guessing becomes an intelligent strategy.) The SAT was a test devised by a particular institution, by a particular kind of person, operating from a particular mindset. It had an ideology, and Kaplan realized that anyone who understood that ideology would have a tremendous advantage.

Critics of the SAT have long made a kind of parlor game of seeing how many questions on the reading-comprehension section (where a passage is followed by a series of multiple-choice questions about its meaning) can be answered without reading the passage. David Owen, in the anti-SAT account "None of the Above," gives the following example, adapted from an actual SAT exam:

1. The main idea of the passage is that:
   A) a constricted view of [this novel] is natural and acceptable

B) a novel should not depict a vanished society
C) a good novel is an intellectual rather than an emotional experience
D) many readers have seen only the comedy [in this novel]
E) [this novel] should be read with sensitivity and an open mind

If you've never seen an SAT before, it might be difficult to guess the right answer. But if, through practice and exposure, you have managed to assimilate the ideology of the SAT — the kind of decent, middlebrow earnestness that permeates the test — it's possible to develop a kind of gut feeling for the right answer, the confidence to predict, in the pressure and rush of examination time, what the SAT is looking for. A is suspiciously postmodern. B is far too dogmatic. C is something that you would never say to an eager, college-bound student. Is it D? Perhaps, but D seems too small a point. It's probably E — and sure enough, it is.

With that in mind, try this question:

2. The author of [this passage] implies that a work of art is properly judged on the basis of its:
   A) universality of human experience truthfully recorded
   B) popularity and critical acclaim in its own age
   C) openness to varied interpretations, including seemingly contradictory ones
   D) avoidance of political and social issues of minor importance
   E) continued popularity through different eras and with different societies

Is it any surprise that the answer is A? Bob Schaeffer, the public education director of the anti-test group FairTest, says that when he got a copy of the latest version of the SAT, the first thing he did was try the reading comprehension section blind. He got twelve out of thirteen questions right.

The math portion of the SAT is perhaps a better example of how coachable the test can be. Here is another question, cited by Owen, from an old SAT:

> In how many different color combinations can 3 balls be painted if each ball is painted one color and there are 3 colors available? (Order is not

considered; e.g. red, blue, red is considered the same combination as red, red, blue.)

A) 4 B) 6 C) 9 D) 10 E) 27

This was, Owen points out, the twenty-fifth question in a twenty-five-question math section. SATs — like virtually all standardized tests — rank their math questions from easiest to hardest. If the hardest questions came first, the theory goes, weaker students would be so intimidated as they began the test that they might throw up their hands in despair. So this is a "hard" question. The second thing to understand about the SAT is that it only really works if good students get the hard questions right and poor students get the hard questions wrong. If anyone can guess or blunder his way into the right answer to a hard question, then the test isn't doing its job. So this is the second clue: the answer to this question must not be something that an average student might blunder into answering correctly. With these two facts in mind, Owen says, don't focus on the question. Just look at the numbers: there are three balls and three colors. The average student is most likely to guess by doing one of three things — adding three and three, multiplying three times three, or, if he is feeling more adventurous, multiplying three by three by three. So six, nine, and twenty-seven are out. That leaves four and ten. Now, he says, read the problem. It can't be four, since anyone can think of more than four combinations. The correct answer must be D, ten.

Does being able to answer that question mean that a student has a greater "aptitude" for math? Of course not. It just means that he had a clever teacher. Kaplan once determined that the testmakers were fond of geometric problems involving the Pythagorean theorem. So an entire generation of Kaplan students were taught "boo, boo, boo, square root of two," to help them remember how the Pythagorean formula applies to an isosceles right triangle. "It was usually not lack of ability," Kaplan writes, "but poor study habits, inadequate instruction, or a combination of the two that jeopardized students' performance." The SAT was not an aptitude test at all.

In proving that the SAT was coachable, Stanley Kaplan did something else, which was of even greater importance. He undermined the use of aptitude tests as a means of social engineering. In the years immediately before and after World War I, for instance, the country's elite colleges faced what became known as "the Jewish problem." They were being inundated with the children of Eastern European Jewish immigrants. These students came from the lower middle class and they disrupted the genteel WASP sensibility that had been so much a part of the Ivy League tradition. They were guilty of "underliving and overworking." In the words of one writer, they "worked far into each night [and] their lessons next morning were letter perfect." They were "socially untrained," one Harvard professor wrote, "and their bodily habits are not good." But how could a college keep Jews out? Columbia University had a policy that the New York State Regents Examinations — the statewide curriculum-based high school graduation examination — could be used as the basis for admission, and the plain truth was that Jews did extraordinarily well on the Regents Exams. One solution was simply to put a quota on the number of Jews, which is what Harvard explored. The other idea, which Columbia followed, was to require applicants to take an aptitude test. According to Herbert Hawkes, the dean of Columbia College during this period, because the typical Jewish student was simply a "grind," who excelled on the Regents Exams because he worked so hard, a test of innate intelligence would put him back in his place. "We have not eliminated boys because they were Jews and do not propose to do so," Hawkes wrote in 1918.

> We have honestly attempted to eliminate the lowest grade of applicant and it turns out that a good many of the low grade men are New York City Jews. It is a fact that boys of foreign parentage who have no background in many cases attempt to educate themselves beyond their intelligence. Their accomplishment is over 100 percent of their ability on account of their tremendous energy and ambition. I do not believe however that a College would do well to admit too many men of low mentality who have ambition but not brains.

Today Hawkes's anti-Semitism seems absurd, but he was by no means the last person to look to aptitude tests as a means of separating ambition from brains. The great selling point of the SAT has always been that it promises to reveal whether the high school senior with a 3.0 GPA is someone who could have done much better if he had been properly educated or someone who is already at the limit of his abilities. We want to know that information because, like Hawkes, we prefer naturals to grinds: we think that people who achieve based on vast reserves of innate ability are somehow more promising and more worthy than those who simply work hard.

But is this distinction real? Some years ago, a group headed by the British psychologist John Sloboda conducted a study of musical talent. The group looked at 256 young musicians, between the ages of ten and sixteen, drawn from elite music academies and public school music programs alike. They interviewed all the students and their parents and recorded how each student did in England's national music-examination system, which, the researchers felt, gave them a relatively objective measure of musical ability. "What we found was that the best predictor of where you were on that scale was the number of hours practiced," Sloboda says. This is, if you think about it, a little hard to believe. We conceive musical ability to be a "talent" — people have an aptitude for music — and so it would make sense that some number of students could excel at the music exam without practicing very much. Yet Sloboda couldn't find any. The kids who scored the best on the test were, on average, practicing *800 percent more* than the kids at the bottom. "People have this idea that there are those who learn better than others, can get further on less effort," Sloboda says. "On average, our data refuted that. Whether you're a dropout or at the best school, where you end up can be predicted by how much you practice."

Sloboda found another striking similarity among the "musical" children. They all had parents who were unusually invested in their musical education. It wasn't necessarily the case that the parents were themselves musicians or musically inclined. It was simply that they wanted their children to be that way. "The parents of the high achievers did things that most parents just don't do," he said. "They didn't simply drop their child at the door of the teacher. They went into the practice room. They took notes on what the teacher said, and when they got home they would say, Remember when your teacher said do this and that. There was a huge amount of time and motivational investment by the parents."

Does this mean that there is no such thing as musical talent? Of course not. Most of those hardworking children with pushy parents aren't going to turn out to be Itzhak Perlmans; some will be second violinists in their community orchestra. The point is that when it comes to a relatively well defined and structured task — like playing an instrument or taking an exam — how hard you work and how supportive your parents are have a lot more to do with success than we ordinarily imagine. Ability cannot be separated from effort. The testmakers never understood that, which is why they thought they could weed out the grinds. But educators increasingly do, and that is why college admissions are now in such upheaval. The Texas state university system, for example, has, since 1997, automatically admitted any student who places in the top 10 percent of his or her high school class — regardless of SAT score. Critics of the policy said that it would open the door to students from marginal schools whose SAT scores would normally have been too low for admission to the University of Texas — and that is exactly what happened. But so what? The "top 10 percenters," as they are known, may have lower SAT scores, but they get excellent grades. In fact, their college GPAs are the equal of students who scored two hundred to three hundred points higher on the SAT. In other words, the determination and hard work that propel someone to the top of his high school class — even in cases where that high school is impoverished — are more important to succeeding in college (and, for that matter, in life) than whatever abstract quality the SAT purports to measure. The importance of the Texas experience cannot be overstated. Here, at last, is an intelligent alternative to affirmative action, a way to find successful minority students without sacrificing academic performance. But we would never have got this far without Stanley Kaplan — without someone first coming along and puncturing the mystique of the SAT. "Acquiring test-taking skills is the same as learning to play the piano or ride a bicycle," Kaplan writes. "It requires practice, practice, practice. Repetition breeds familiarity. Familiarity breeds confidence." In this, as in so many things, the grind *was* the natural.

To read Kaplan's memoir is to be struck by what a representative figure he was in the postwar sociological miracle that was Jewish Brooklyn. This is the lower-middle-class, second- and third-generation immigrant world, stretching from Prospect Park to Sheepshead Bay, that ended up peopling the upper reaches of American professional life. Thousands of students from those neighborhoods made their way through Kaplan's classroom in the fifties and sixties, many along what Kaplan calls the "heavily traveled path" from Brooklyn to Cornell, Yale, and the University of Michigan. Kaplan writes of one student who increased his score by 340 points and ended up with a Ph.D. and a position as a scientist at Xerox. "Debbie" improved her SAT by 500 points, got into the University of Chicago, and earned a Ph.D. in clinical psychology. Arthur Levine, the president of Teachers College at Columbia University, raised his SATs by 282 points, "making it possible," he writes on the book's jacket, "for me to attend a better university than I ever would have imagined." Charles Schumer, the senior senator from New York, studied while he worked the mimeograph machine in Kaplan's office, and ended up with close to a perfect 1600.

These students faced a system designed to thwart the hard worker, and what did they do? They got together with their pushy parents and outworked it. Kaplan says that he knew a "strapping athlete who became physically ill before taking the SAT because his mother was so demanding." There was the mother who called him to say, "Mr. Kaplan, I think I'm going to commit suicide. My son made only one thousand on the SAT." "One mother wanted her straight-A son to have an extra edge, so she brought him to my basement for years for private tutoring in basic subjects," Kaplan recalls. "He was extremely bright and today is one of the country's most successful ophthalmologists." Another student was "so nervous that his mother accompanied him to class armed with a supply of terry-cloth towels. She stood outside the classroom, and when he emerged from our class sessions dripping in sweat, she wiped him dry and then nudged him back into the classroom." Then, of course, there was the formidable four-foot-eight figure of Ericka Kaplan, granddaughter of the chief rabbi of the synagogue of Prague. "My mother was a perfectionist whether she was keeping the company books or setting the dinner table," Kaplan writes, still in her thrall today. "She was my best cheerleader, the reason I performed so well, and I constantly strove to please her." What chance did even the most artfully constructed SAT have against the mothers of Brooklyn?

Stanley Kaplan graduated number two in his class at City College and won the school's Award for Excellence in Natural Sciences. He wanted to be a doctor, and he applied to five medical schools, confident that he would be accepted. To his shock, he was rejected by every single one. Medical schools did not take public colleges like City College seriously. More important, in the forties there was a limit to how many Jews they were willing to accept. "The term *meritocracy* — or success based on merit rather than heritage, wealth, or social status — wasn't even coined yet," Kaplan writes, "and the methods of selecting students based on talent, not privilege, were still evolving."

That's why Stanley Kaplan was always pained by those who thought that what went on in his basement was somehow subversive. He loved the SAT. He thought that the test gave people like him the best chance of overcoming discrimination. As he saw it, he was simply giving the middle-class students of Brooklyn the same shot at a bright future that their counterparts in the private schools of Manhattan had. In 1983, after years of hostility, the College Board invited him to speak at its annual convention. It was one of the highlights of Kaplan's life. "Never, in my wildest dreams," he began, "did I ever think I'd be speaking to you here today."

The truth is, however, that Stanley Kaplan was wrong. What he did in his basement *was* subversive. The SAT was designed as an abstract intellectual tool. It never occurred to its makers that aptitude was a social matter: that what people were capable of was affected by what they knew, and what they knew was affected by what they were taught, and what they were taught was affected by the industry of their teachers and parents. And if what the SAT was measuring, in no small part, was the industry of teachers and parents, then what did it mean? Stanley Kaplan may have loved the SAT. But when he stood up and recited "boo, boo, boo, square root of two," he killed it.

GARY GREENBERG

As Good as Dead

FROM *The New Yorker*

JUST AFTER a fourteen-year-old boy named Nicholas Breach learned that a tumor on his brain stem would be fatal, he told his parents, Rick and Kim Breach, that he wanted to be an organ donor. They respected his decision, and so did the boy's medical team at the Children's Hospital of Philadelphia. Bernadette Foley, Nick's social worker there, said that the decision reflected a "maturity and sensitivity" and a wish to help others — something Nick had shown throughout his eight-year battle with recurrent tumors. "I've never been to a meeting like this one," Foley said. "The peace that came over the family and Nick was remarkable, and once it was out that this was the end, and the decision was made about organ donation, Nick said he was happy. They all seemed to be happy." The decision was redemptive, she said. "In a way, it gave some meaning to his life."

By the time I met Nick, he was confined to a hospital bed that had been set up in the living room of the Breaches' house, a brick bungalow outside Harrisburg. It was difficult for him to speak, and we chatted only briefly — about his dog, Sarah; his brother, Nathan; and his hope that his heart, lungs, liver, kidneys, and pancreas might enable other people to live — and then he dozed off.

As Nick slept, his parents told me that, amid their other worries, they had run into unexpected problems with the donation. Nick had wanted to die at home, with only palliative care, but organ donation is a high-tech affair. In most cases, the donor is someone with brain damage so severe that he requires a respirator to breathe, even though his heart continues to work on its own. A neurologist determines that the patient's brain has been irreversibly and totally destroyed, and on this basis pronounces him dead. This condition is known as brain death. If the patient's family has consented to donation, he is left on the respirator, which, along with his still-beating heart, keeps his organs viable for transplant until they can be harvested. The Breaches accepted that Nick would now have to be hospitalized at the very end, but their insurance company balked at the change in plan — and the added expense — reminding them that they had already elected basic hospice care. Only after the family's state legislator and the regional organ-procurement organization got involved did the insurance company agree to pay. A plan was devised to keep Nick at home until the last possible moment and then to transport him to a hospital, where an informal protocol had been set up to help him become an organ donor.

Even with the logistical and financial arrangements in place, there was no guarantee that Nick would meet the criteria for brain death. Because the tumor was on his brain stem, which controls core physiological processes like breathing and body temperature, it was very likely that Nick's higher brain — the thinking part — would remain active until he died from respiratory or organ failure. (His oncologist told me, "In his condition, what happens is the body goes. He's a consciousness trapped inside.") This would probably rule him out as a donor.

When I spoke to Nick's parents, they still had trouble with the notion that, to become a donor, it was not enough for their son to die with his body more or less intact. He would have to have the right kind of death, with the systems in his body shutting down in a particular order. "I'm so confused about this part of it," his mother said. "I don't understand why, if his heart stops beating, they can't put him back on a respirator." Rick, too, was confused about the moment at which "the plug will be pulled." In reality, there is no moment when the plug is pulled; to keep the organs viable, the respirator is left operating — and the heart keeps beating — until the surgeon removes the organs.

Confusion about the concept of brain death is not unusual, even among the transplant professionals, surgeons, neurologists, and bioethicists who grapple with it regularly. Brain death is confusing because it's an artificial distinction constructed, more than thirty years ago, on a conceptual foundation that is unsound. Recently, some physicians have begun to suggest that brain-dead patients aren't really dead at all — that the concept is just the medical profession's way of dodging ethical questions about a practice that saves more than fifteen thousand lives a year.

From the beginning, transplant practice has been governed by a simple, unwritten rule: no matter how extreme the circumstances, no matter how ill or injured the potential donor, he must die of some other cause before his vital organs can be removed; it would never be acceptable to kill someone for his organs. But, ideally, a donor would be alive at the time his organs were harvested, because as soon as the flow of oxygenated blood stops, a process called warm ischemia quickly begins to ruin them. By the 1960s, as doctors began to perfect techniques for transplanting livers and hearts, the medical establishment faced a paradox: the need for both a living body and a dead donor.

The profession was also struggling with questions posed by another new technology: respirators. These machines had become a fixture in hospitals in the 1950s, and at first their main purpose was to help children with polio breathe until they regained their strength. Doctors began to use them for patients with devastating brain injuries — the kind brought on by severe trauma or loss of oxygen as a result of stroke or cardiac arrest. Some of these people recovered sufficiently to be removed from the machines, but others lingered, unable to breathe on their own, inert and unresponsive even to the most noxious stimulus, and without any detectable electrical brain activity, until their hearts gave out — often a matter of hours, but sometimes of days or even weeks.

Physicians wondered what to do with these patients, whether removing the machines would be murder or mercy killing or simply a matter of letting nature take its course. At the same time, some noticed that the patients were perfect sources of viable organs for transplant, at least as long as their hearts kept beating. And then, in 1967, a Harvard anesthesiologist named Henry K. Beecher asked the dean of the medical school to form a committee to explore the issues of artificial life support and organ donation, which he believed were related. The Harvard committee, which Beecher chaired, included ten physicians, a lawyer, and a historian, and its report was published the following year in the *Journal of the American Medical Association*. "Responsible medical opinion," it announced, "is ready to adopt new criteria for pronouncing death to have occurred in an individual sustaining irreversible coma as a result of permanent brain damage." Heartbeat or no, the committee declared, patients whose brains no longer functioned and who had no prospect of recovering were not lingering but were already dead — brain dead.

This physician-assisted redefinition of death meant that removing life-support machinery from these patients was no longer ethically suspect. And, by creating a class of dead people whose hearts were still beating, the Harvard committee gave transplant surgeons a new potential supply of organs. In the 1970s, however, only twenty-seven states adopted brain death as a legal definition of death. Theoretically, this meant that someone who had been declared dead in North Carolina could be resurrected by transferring him to a hospital in South Carolina. Practically, it meant that a doctor procuring organs from a brain-dead person was not equally protected in all jurisdictions from the charge that he was killing his patient.

In 1980, a commission appointed by President Carter began to look at medical ethical questions, which included finding a definition of death that could serve as a model for state laws. The commission recommended that doctors be given the power to declare people dead based on the neurological criteria suggested by the Harvard committee. Eventually, this recommendation was accepted in all fifty states.

The commission also wanted to convince the public that brain death was not just a legal fiction but the description of a biological truth. Two rationales were considered. In one, called the "higher-brain" formulation, a brain-dead person is alleged to be dead because his neocortex, the seat of consciousness, has been destroyed. He has thus lost the ability to think and feel — the capacity for personhood — that makes us who we are, and our lives worth living. But such "quality of life" criteria, the commission noted, raised uncomfortable ethical and political questions about the treatment of senile patients and how society valued the lives of the mentally impaired.

Instead, the commission chose to rely on what it called the "whole-brain" formulation. The brain, it was argued, directed and gave order and purpose to the different mechanical functions of our bodies. If both the neocortex and the brain stem (which regulates core physiological processes, such as breathing) stopped working, a person could be pronounced dead — not just because consciousness has disappeared but because, without the brain, nothing connects: there is no internal harmony, and the body no longer exists as an integrated whole.

When Nick Breach decided to become a donor, one of his first questions was whether he would be dead when his organs were taken. His parents told him that he would be, and, in a way, they saw this as one of the few things they could be sure about. Rick and Kim were more troubled by their son's next concern, that he might be taken from them prematurely. They began a vigil that took on a strange dual nature: keeping Nick company, making him comfortable, spending as much time as possible with him, and, at the same time, monitoring him for the signs — whatever they might be — that death had come so close that it was time to get him to the hospital so that he could become an organ donor.

The organ-procurement agency that worked with the Breaches during those months was called Gift of Life. In 2000, Gift of Life, which is based in Philadelphia and has a staff of a hundred, helped manage more than 800 organ donations at 162 member hospitals in Pennsylvania, New Jersey, and Delaware — 5 percent of the total organs removed in the country.

The agency's mission is to "positively predispose all members of the community to organ and tissue donation so that donation is viewed as a fundamental human responsibility." Public-service ads, a pamphlet featuring Michael Jordan, and bumper stickers that say DON'T TAKE YOUR ORGANS TO HEAVEN — HEAVEN KNOWS WE NEED THEM HERE are all promoting an attitude about how, as Howard M. Nathan, the bearded, energetic forty-seven-year old who heads Gift of Life, put it, "society should feel about this subject." Because of the drama and human interest of Nick Breach's case, the agency was naturally eager to publicize it: "Here's a young man who is awake and aware, contemplating his death, and he becomes a donor," Kevin Sparkman, the agency's director of community relations, explained. "What a great example of what we want families to do!"

When a person is identified as a potential organ donor — generally, when he is about to be pronounced brain dead — Gift of Life dispatches a transplant coordinator to the hospital to try to obtain the family's consent. (An organ-donor card is merely an indication of a patient's wishes; the family has the final word.) "The first thing we do is ensure that the family understands and acknowledges that their loved one is dead," Linda Herzog, a senior hospital-services coordinator, told me.

Consent rates are tied directly to knowledge of brain death: families who think that donation is actually going to kill the patient refuse more often than families who believe that their relative is already dead. This is not as straightforward as it may seem, largely because of the lifelike appearance of the brain dead, whose skin is still warm to the touch and who are known within the industry as "heart-beating cadavers." Gift of Life has developed a program that trains hospital staffs to explain the phenomenon to families. I watched in a darkened conference room as Herzog reviewed the program for two transplant coordinators, who were scheduled to present it later that afternoon in a Philadelphia hospital.

Using slides, Herzog ran through the process by which brain death is established. A neurologist performs a series of tests at the bedside — checking for such things as pupillary reflexes, response to pain, and the ability to breathe spontaneously. (If the patient is entirely unresponsive during two such examinations, the doctor concludes that his whole brain — cortex and brain stem — has been destroyed.) This is not a terribly sophisticated procedure, but it's far more complicated than, say, ascertaining that a person has no pulse, and far less self-evident. Even when the tests are conducted or reenacted in front of family members, they often rely on their intuitions and insist that the patient is still alive. This failure to accept the truth is a function of denial, Herzog said, and she went on to note, with some dismay, that even highly trained professionals who fully accept the concept sometimes talk to brain-dead patients.

"It took us years to get the public to understand what brain death was," Nathan said. "We had to train people in how to talk about it. Not that they're brain dead, but they're dead: 'What you see is the machine artificially keeping the body alive . . .'" He stopped and pointed to my notebook. "No, don't even use that. Say 'keeping the organs functioning.'"

Virtually every expert I spoke with about brain death was tripped up by its semantic trickiness. "Even I get this wrong," said one physician and bioethicist who has written extensively on the subject, after making a similar slip. Stuart Youngner, the director of the Center for Biomedical Ethics at Case Western Reserve University, thinks that the need for linguistic vigilance indicates a problem with the concept itself. "The organ-procurement people and transplant activists say you've got to stop saying things like that because that promulgates the idea that the patients are not really dead. The language is a symptom not of stupidity but of how people experience these 'dead' people — as not exactly dead."

Last year, I went to Havana for the Third International Symposium on Coma and Death, a conference held every four years and attended primarily by neurologists and bioethicists, joined by lawyers, anthropologists, and members of the clergy. At one session, I watched as a videotape of a recumbent adolescent boy, his feet toward the camera, his legs bowed, almost froglike, played on a television monitor in a corner of the room. He wore shorts, and there were two tubes entering his body, one in his abdomen, the other in his throat. The boy's chest rose and fell to the whir and click of the respirator, but otherwise he was perfectly still.

On the tape, a trim, balding man named Alan Shewmon, a pediatric neurologist at UCLA, stood near the bed and conducted a medical examination. He looked into the boy's eyes, shook maracas next to his head, inserted a swab in a nostril, dropped cold water into the ears and lemon on the tongue, pinched and palpated and inspected. None of these actions drew a response from the boy, whom I will call Matthew.

Shewmon was also standing next to the monitor in Havana, offering additional commentary. He has been thinking about death for most of his career. A practicing Catholic, he has made contesting the concept of brain death a specialty, and has served on a Pontifical Academy of Sciences task force on the subject. Shewmon's inquiry has led him from the higher-brain rationale through the whole-brain rationale to his current position: a strong conviction that brain death, while a severe disability, even severe enough to warrant discontinuing life support, is not truly death.

Although Matthew didn't seem dead, it was hard to think of him as alive. On the monitor, a nurse removed the upper tube, suctioned the small hole in the boy's throat, noted that he did not cough, and continued the routine of the exam. Then something different happened: some ice water trickled onto the boy's shoulder, and it twitched. And though the screen was too small to see this, Shewmon told us that Matthew sprouted goose bumps, that his flesh was mottling and flushing with the stress of the exam. He was showing signs, that is, of precisely the kind of systemic functioning that the brain dead would generally be expected to lack.

In the video, Shewmon lifted Matthew's arm by the wrist, and the hand sprang to life with a small spasm. A woman's voice — Matthew's mother, we learned — said, "When he knows what you're going to do, he stops that." Shewmon described what was going on in medical terms — clonus, an involuntary contraction and release of nerves. He was making his main point: that this boy — who at age four was struck with meningitis that swelled his brain and split his skull, who would probably have been pronounced brain dead had he not been too young under the statutes of the time, whose mother refused to discontinue life support and ultimately took her son home on a ventilator and a feeding tube, who had persisted in this twilight condition for thirteen years, healing from wounds and illness, growing — was alive. Not by virtue of intention or will, as his mother has implied, but because he had maintained a somatic integrated unity — the internal harmony, and the overarching coordination of his body's functions — which, if the whole-brain rationale is correct, he simply should not have been able to do.

After the presentation ended, I spoke to Ronald Cranford, a professor of neurology and bioethics at the University of Minnesota, who is one of Shewmon's critics. He argued that Matthew's case was only an unusually prolonged example of the normal course brain death takes. "Any patient you keep alive, or dead, longer than a few days will develop spinal-cord reflexes," he said, recalling a case in which the doctor said, "Yes, she's been getting better ever since she died."

In a question-and-answer session with Shewmon the next day, after an address in which he drew parallels between the brain dead and people who are conscious but have been paralyzed by injuries to the upper spinal cord, no one really took issue with his science. At the same time, none of the physicians would accept what

Shewmon was really saying: that the brain dead are not dead. "The main philosophical question is, Is this a body or is this a person?" said Calixto Machado, the Cuban neurologist who organized the symposium. Fred Plum, the chairman emeritus of the Department of Neurology at Cornell University's Weill Medical College, had positioned himself directly in front of the podium for the talk, and shot his hand in the air as soon as Shewmon was finished. "This is anti-Darwinism," Plum said. "The brain is the person, the evolved person, not the machine person. Consciousness is the ultimate. We are not one living cell. We are the evolution of a very large group of systems into the awareness of self and the environment, and that is the production of the civilization in which any of us lives."

Shewmon had laid a trap for his audience, he later told me. He had hoped to break down the pretense that anyone subscribed to the whole-brain rationale. He wanted to show that the higher-brain rationale, which holds that living without consciousness is not really living — and which the president's commission rejected because it raised questions about quality of life which science can never settle — was the sub-rosa justification for deciding to call a brain-dead person dead. He wanted to make it clear that these doctors were not making a straightforward medical judgment but, rather, a moral judgment that people like Matthew were so devastated that they had lost their claim on existence. And, at least in his own view, the comments he'd provoked meant that he had succeeded.

The neurologist James Bernat, a professor at Dartmouth Medical School and the author of the chapters on brain death in several neurology textbooks, is one of the defenders of the whole-brain concept. Like Shewmon, Bernat served on the Pontifical Academy of Sciences task force. And, last August, his position appeared to prevail when Pope John Paul II, speaking before an international transplantation congress, said that "the complete and irreversible cessation of all brain activity, if rigorously applied," along with the family's consent, gave a "moral right" to remove organs for transplant — thus resolving an ambiguity in the Church as to whether Catholics should become donors. (Orthodox Jewish and other theologians continue to debate whether a brain-dead person is truly dead.) But even Bernat sees the problem he's up against. "Brain death was accepted before it was conceptually sound," he told me on the telephone from his office in New Hampshire. He readily admits that no one has yet explained scientifically why the destruction of the brain is the death of the person, rather than an extreme injury. "I'm being driven by an intuition that the brain-centered concept of death is sound," he said. "Death is a biological function. Death is an event."

Stuart Youngner, of Case Western, however, rails against what he sees as bad faith in the way brain death came to be defined. Youngner, a white-bearded, avuncular fifty-six-year-old, calls the Harvard committee's work "conceptual gerrymandering," a redrawing of the line between life and death which was determined by something other than science. "What if the Harvard committee, instead of saying, 'Let's call them dead,' had said, 'Let's have a discussion in our society about whether there are circumstances in which people's organs can be taken without sacrificing freedom, without harming people.' Would it be better?"

The problem, as Youngner sees it, is that the veneer of scientific truth attached to the concept of brain death conceals the fact that the lives of brain-dead people have ended only by virtue of what amounts to a social agreement. According to Youngner, this means that the brain dead are really just "as good as dead," but, he is quick to add, this doesn't mean that they shouldn't be organ donors. Instead, he suggests that "as good as dead" be recognized as a special status, one that many people, brain dead or not, may achieve at the end of life. "I'm willing to point out the ambiguities and inconsistencies in the notion, and I actually think that acknowledging them may in the long run be better," he told me.

During the last decade, Youngner and other doctors and ethicists have developed protocols to allow critically ill or injured people who have no hope of recovery, but who are unlikely to become brain dead, to donate their organs after they have been declared dead by the traditional cardiopulmonary criteria. This procedure, which is known as non-heart-beating-cadaver donation and requires extremely rapid intervention and newly developed techniques, may make it possible to salvage viable organs in a wider range of cases.

As it happened, Nick Breach was a candidate for this procedure. If he was brought to the hospital, placed on a respirator, and then languished, without ever meeting the criteria for brain death — a likely scenario, given the course of his disease — only a tight orchestration of his death could conceivably give him a chance of becoming a donor. According to Gift of Life's protocol, Nick's parents would first have to decide to remove life support. Nick would then be taken to an operating room, where he would be taken off the ventilator, and the doctors would wait for his heart to stop. If that took more than an hour, warm ischemia would set in (as his breathing would be too compromised to supply oxygen to his organs), the donation would be aborted, and Nick would be returned to a hospital room to die. But if cardiac arrest came in time, a five-minute count would begin, at the end of which Nick would be declared dead. A transplant team standing by in an anteroom would immediately harvest his organs and rush them to their recipients. (Even with this alternative, the window for success was fairly narrow. "All we're trying to do," Howard Nathan acknowledged, "is give it a shot.")

Non-heart-beating protocols have the potential to increase donation by as much as 25 percent. But, as Youngner points out, the five-minute waiting period (it ranges from two minutes in some protocols to ten minutes in others) is really just a decent interval, a more or less arbitrary marker of the passage from life to death, whose significance is far more symbolic than scientific.

Robert Truog, a professor of medical ethics and anesthesiology at Harvard Medical School, is even more critical of the protocol. "Non-heart-beating protocols are a dance we do so that people can comply with the dead-donor rule," he told me. "It seems silly that we hang on to this facade. It's a bizarre way of practice, to be unwilling to say what you are doing" — that is, identifying a person as an organ donor when he is still alive and then declaring him dead by a process tailored to keep up appearances and which, in the bargain, might not best meet the requirements of transplant. In Truog's view, a better approach would be to remove these patients' organs while they are still on life support, as is done with brain-dead donors. "If they have detectable brain activity, then they should be given anesthetic," he said, but there is no reason to continue to conceal what is happening by waiting for their hearts to stop beating.

Abandoning the dead-donor convention — which is an inevitable consequence of Youngner's and Truog's positions — may, however, cause other problems. It awakens the same sort of fears that

Nick Breach himself had about the premature removal of his organs. It raises vexing legal questions, because, as Truog bluntly told me, without the rule "taking organs is a form of killing" — killing that he thinks is justified, and that Youngner and others would argue is already happening. He added that repealing the rule risks "making physicians seem like a bunch of vultures."

In return, Truog points out, patients would gain more control over the end of their lives: they would no longer have to wait until they crossed over that gerrymandered border and, instead, could specify at what point they would like to be declared dead so that they could donate their organs. This, however, might not be adequate consolation for those who fear that the need for organs might create a perverse incentive for doctors to give up on them, after weighing their lives against those of others who may be more worthy or less damaged. Youngner expressed reservations about how his position would sound to other doctors and, most important, to potential donors. "I think that stridently advocating the abandonment of the dead-donor rule would be a mistake," he said. He worried, he told me, that religious conservatives and others might "seize on it as a violation of the right to life," thus turning transplant into another medical practice — like abortion or fetal stem-cell research — that's bogged down in intractable political wrangling.

As Nick Breach thought about his death, he made some additional last wishes that were easier to satisfy than his desire to become a donor: Ronald McDonald came to visit; so did Weird Al Yankovic, one of Nick's idols. When Yankovic pulled up to the Breaches' house in a bus, the neighbors moved their cars to accommodate him. Yankovic came inside and sat for a while with Nick, who was bedridden by then. Nick told him, "I really love all your CDs, Weird Al."

Six days later, at 11:45 P.M., Nick stopped breathing. Rick, who was taking his turn by the bedside, summoned Kim, who called an ambulance and began to administer CPR. The plan was to revive Nick so that he could be brought to a hospital and placed on a ventilator. But his mother's efforts, and those of the paramedics in the ambulance and the staff in the emergency room, failed. Nick's heart had stopped too soon, and ischemia had set in. In the end, the only organs he was able to donate were his eyes.

It is tempting to wish that death weren't so complicated. Had Nick and his parents realized how alive he still needed to be in order to donate his vital organs successfully, they could have been given an honest choice between having Nick remain at home until the end and giving up on his goal of becoming a donor, and going to the hospital much earlier and staying until he could be declared "as good as dead."

Over and over again at the conference in Havana, I heard ambivalence and anxiety about "the public" knowing what doctors already know. "These things ought to be worked out in the medical profession, to some extent, before you go to the public," Shewmon told me. "Because if you go public right away, it could just put the kibosh on the whole thing, because people get hysterical and misunderstand things." He paused and looked at me. "These are complex issues. You can't expect the public to understand these things in sound bites, which is what they usually get. So I'm reluctant to talk to reporters about this stuff."

During a break between sessions, I got into a conversation with a philosopher. He told me that he had been talking about this subject with a colleague, and that they'd found themselves calling brain death a "noble lie." Later, as the conference reconvened, I asked him if we could talk some more about that idea. He was visibly upset. "Listen, I'm not sure about that comment," he said. "It's inflammatory. It's too strong." Among his concerns, he explained, was the possibility that his words might discourage people from becoming organ donors.

It may be too much to say that the concept of brain death is an outright lie, but it is certainly less than the truth. Like many of technology's sublime achievements, organ transplant, for all its promise, also has an unavoidable aspect of horror — the horror of rendering a human being into raw materials, of turning death into life, of harvesting organs from an undead boy. Should a practice, however noble, be able to hold truth hostage? Perhaps the medical profession should embrace the obvious: to be an organ donor is to choose a particular way to finish our dying, at the hands of a surgeon, after some uncertain border has been crossed — a line that will change with time and circumstance, and one that science will never be able to draw with precision.

GORDON GRICE

Is That a Mountain Lion in Your Backyard?

FROM *Discover*

WHEN HE KICKED IT, the cow's head flipped completely over. It lay flat, the desiccated and cracked pink nose pointing opposite to the way it had lain before. The spine was broken.

The man — I'll call him Harris because, although he feels he's done nothing wrong, he's not eager for authorities to know his real name — would lose several head of cattle that year, as would some of his neighbors. Coyotes were the usual culprits on these little ranches just north of the Colorado–New Mexico border. But Harris could see this kill was not the work of coyotes. He knew mountain lions had come back to this part of the country.

All the mountain lion kills that year were similar. The carcass, which might weigh hundreds of pounds, would turn up several yards from the kill site, sometimes with no drag marks between the two places. A 140-pound cat can lift and carry three times its own weight in its mouth. (At a University of California at Davis research facility, a mountain lion once killed a 130-pound deer and carried it over an eight-foot-high chainlink fence.) A cat would open the belly first and set aside the stomach — the vegetable matter inside apparently considered unpalatable. It would eat the heart, liver, and lungs. Often it would cover the carcass with dirt and vegetation and save it for later, but that wouldn't always fool the crows and foxes and badgers that would come to scavenge. When a lion would finish a carcass, it would eat everything but the fur.

*

Four hundred years ago the mountain lion ranged over all of the Americas except the extreme north. It was known by many names, including mountain screamer and indian devil. The appellations still in common use include cougar, puma, and panther. It was a top predator, more formidable than the wolf or the coyote or the bobcat, its superior speed leaving it little to fear from the bear or the jaguar. When Western civilization arrived, the invaders set out to exterminate their rivals — not just the human ones, but the animals as well. The Massachusetts Bay Colony offered bounties as early as 1630 to encourage the slaughter of predators. In Virginia, where bounties were instituted in 1632, native tribes were required to pay an annual tribute of wolf hides to the colonial government. Bounties continued well into the twentieth century. Another practice with a colonial pedigree was the roundup or drive, in which coyotes, rattlesnakes, and other predators were killed by the hundreds.

Even now, most farm households keep guns for dealing with predators that might take livestock. The prevailing ethic before environmentalism wasn't simply to kill in defense of life or property; it was to kill any predator with the potential to do anyone harm later on. I still know men who shoot at any coyote they happen to spot. Some make a point of running over coyotes crossing the road. These are ordinary men — that's what they're supposed to do in the culture that produced them.

The jaguar made an early exit from the United States, hunted so aggressively that most people don't know it ever existed north of Mexico. The gray wolf has come perilously close to disappearing in the contiguous states. The alligator neared extinction before management brought it back. Of North America's large predators, only the coyote has prospered since the arrival of Europeans, expanding its range and its population. Its success stems partly from the decimation of the wolf, its slightly less adaptable competitor.

The mountain lion is a strange case. It has almost vanished from the eastern United States — although an occasional lion sighting is reported but never confirmed up in New England, and Florida has a small population. But mountain lions still inhabit the western half of North America, and in the past decade their populations have risen dramatically. Some have blamed the increase on a lack of hunting. Each state has its own hunting regulations, but in general mountain lions have enjoyed greater protection in the last two or three decades. In California, where sport hunting of mountain lions has been illegal since 1972, the population has boomed until recently, with more ranchers losing more livestock to the cats, and people encountering more of them in backyards. But decreased hunting isn't a complete answer. The mountain lion population has also boomed in Montana, where hunting remains popular.

Even at its lowest ebb, the lion population has never been in danger of dying out. It can live in deserts, swamps, and rain forests, and on mountains and coastal plains. And because it is stealthy and generally solitary, it can live near humans undetected. It prefers to hunt at dawn and dusk but adapts to working by day or night. Like its near cousin, the domestic cat, it is a generalist predator. It often preys on deer, elk, and other large herbivores but is also comfortable eating mice and squirrels, even grasshoppers. Occasionally a mountain lion takes a small alligator.

The animal's physical abilities are impressive. It can see in the dark. It can swim and climb trees. It can fall fifty feet and land on its feet, unharmed. It can execute twenty-five-foot leaps from a dead stop, and with a running start can leap almost forty feet. It has been known to clear ten-foot fences and to leap from the ground onto the back of a person on horseback.

Its usual method of killing large prey is to stalk to within pouncing range, spring onto the victim's back, and drive its fangs into the neck, severing the spine for a quick kill. But it is adaptable in this regard as well: with a large animal, such as a mature elk, the cat will sometimes opt for a strangling bite to the windpipe. Mountain lions have also been known to hamstring large animals on the run.

It's often said that people are the only animals that kill for pleasure, but probably no cat-lover has ever made this claim. Domestic cats stalk more often than they attack, attack more often than they kill, and kill more than they eat. You can explain this as practice or simply as fun. Mountain lions seem to have the same idea of fun. One of them killed thirty sheep in one night.

A few mornings after the discovery of the cow's carcass, the ranch's best saddle horse came in wounded. Parallel claw marks ran horizontally along its flank, like a musical staff drawn in blood. Harris searched the spot where the horse had pastured the night before.

He found a few stray pugs, or footprints, each showing four toes and the M-shaped ball of the foot. The pugs showed no evidence of claws — mountain lions can retract their claws.

Next a gravid cow was missing. When Harris found her in a narrow canyon, she had already given birth. The calf was nowhere to be found. Two more weeks passed. This time an older Hereford calf was missing. Everyone agreed there were fewer deer than usual. Some said this explained why the carnivores were troubling so much livestock; others said the unusually numerous carnivores had killed off the deer.

Harris never found the calf, but he watched the crows until their gathering showed him the killing place. When he rode up he found plenty of crows on the ground, pecking at small objects in the grass. When he dismounted they turned their heads to watch him with their indifferent blue-black eyes. They looked as if they might hold their ground, but when Harris was a few steps away they all simultaneously stretched their glittering wings and leaned forward into flight. He examined the ground. Tufts of the calf's red and white fur lay everywhere. A mountain lion's tongue is studded with spikes that can flense cartilage from bone or hair from hide. A few slivers and chips of bloody bone lay scattered among the fur. Nothing else remained of the calf.

I came by my interest in mountain lions in the spring of 1993. A dozen others and I were staying at a ranch, where we were guests at a wedding. Some of us went out for a ride. Our horses' shoes clapped against the steep granite as we worked our way around the mountain's shoulder. Then we were into a stretch of open field. My horse, a big, unruly bay, trudged through a clattering pile of bones. I reined him in and asked Virgil, the wrangler, about the carcass. We dismounted.

Virgil handed me the skull. It was about as long as my hand, equipped with broad molars for chewing vegetation.

"Pronghorn antelope," Virgil said. "They don't have front teeth. They use their lips to pull in grass and leaves."

The horns themselves were gone. We looked the bones over to see if we could figure out what had killed the pronghorn. We found plenty of marks that may have been laden with meaning, but neither of us knew how to read the bones.

"These look like something chewed them," I said, holding up a femur and an uncertain fragment.

"Could be," Virgil said. "We've got coyotes and mountain lions. Of course, it could have dropped dead of a disease and then anything could chew on it. God knows how long it's been here."

The skull still had some soft gristle and a flap of hide attached. It had the heft of something not yet empty. I didn't think it was too old. I tied it to my saddle to take back to the ranch house. A pair of sluggish gray bullets emerged from an eye socket and crawled down opposite sides of the nose — carrion beetles.

As we walked to the bunkhouse every morning, led by the smell of bacon, we had to pass the pigpen. One day I stopped to lean on the rails and look the pigs over. There were five, all patched with brown and white except for one plump pink hog. I wondered what pigs think of the smell of bacon.

Virgil came up and leaned on the rail next to me. I declined the bent Marlboro he offered. I kicked idly at a fresh piece of lumber that stood out among the weather-beaten planks and posts.

"I put that on last week," Virgil said. "A mountain lion tore the old board off."

The story that emerged was this: Virgil had awakened to the screams of the pink hog. The mountain lion had it by the hind leg and was trying to drag it through the break in the fence. Virgil fired a shotgun in the air to scare the cat off. I could see the deep, black seam of a healing wound on the hog's leg. I asked whether the cat had been back since.

"Not up here close to the house," he said. He had gone fishing at a stock pond two days earlier. When his horse started acting "spooky," he packed his gear and headed for the ranch house. He returned to the pond later that afternoon. The mountain lion's tracks led down to the fallen log at the water's edge where he had sat fishing.

Near dusk Virgil asked me to help him drive a few head of cattle into their evening pasture. The cattle knew the drill; all we had to do was keep them moving. We did it on foot.

We were walking a dirt road, the cattle hustling ahead of us. On our left was a fenced pasture; on our right, heavy brush. The road changed abruptly from hard-packed dirt to a patch where frequent runoff from a hill had left soft, smooth undulations of dirt. That's where I spotted the pugs of the mountain lion. They dappled the road in a straight line for several yards, obscured in two places where cattle had crossed them. A good rain had fallen about two hours earlier, and we thought the tracks must have been made since then.

Soon we had the cattle in their pasture. Virgil wrestled the gate shut; it was broken, so the process took him a few minutes. Suddenly we both looked toward the brush, then at each other, then back at the brush. I scanned the bank of bushes and ditch-grass and tangled trees.

Virgil whispered a long string of profanities. He told me later he had heard something at that moment, a subtle click that might have been the breaking of a twig. I'm pretty sure I didn't hear anything. I just suddenly got a cold feeling in my scalp, and I knew I was being watched. We started toward the ranch house, Virgil cursing steadily. We walked slowly. I puffed myself up to look large. Virgil was smaller than I and would make a more inviting target, I imagined. I noticed he kept me between himself and the brush. "Wish I had my damn shotgun," he said.

We stopped simultaneously. No signal passed between us, but we must have been thinking the same thing. A thick clump of brush jutted into the road ahead of us, and neither of us wanted to go near it. I stomped on a branch that lay in the road, breaking off a manageable truncheon. Virgil picked up a chunk of sandstone. We walked past the bush, and suddenly we were talking about the weather in loud, angry voices, agreeing that it was nice but a little damp in tones that suggested we were planning to kill each other.

We could see the ranch house up the road. Soon we could see our friends lounging on the veranda. We walked slowly, taking turns proclaiming the damn niceness of the weather over our shoulders.

A conversation about politics drifted down from the veranda; someone quipped and several laughed. Why couldn't they shout down the road to us? Or decide to meet us halfway? Finally we were in the yard and away from the brush.

There is no exciting finish to the story. We were safe. Our friends told us we were the victims of our imagination. The next day I followed our tracks along the road to the evening pasture. A fresh set of pugmarks led toward the house. They ran between Virgil's prints and mine, and occasionally turned a circle before rejoining our path. One pugmark fell neatly within the spade-shaped impression of my left boot, four blunt toes and a trapezoidal foot pad deepening the dent of my print.

I took the pronghorn skull home when I left the ranch. I had to throw it out after a couple of days when it began to smell like bad chicken broth.

If you ask people about any large predator, you will soon encounter two contradictory myths. The first holds that the animal is vicious, a frequent killer of human beings or at least of defenseless lambs. The stereotype of the wolf, for example, as an insatiable killer of livestock and sometimes of people has long been used as an excuse to slaughter the animals.

The other myth claims animals have a "natural fear" of people and will attack only with a special excuse — danger to their cubs, rabies, starvation. The idea that we arouse a universal fear implies we're fundamentally different from the rest of the animal kingdom — an arrogant assumption at best.

The truth varies with the species in question. In North America, wolves fit the stereotype of animals with a natural fear of humanity. They'll go miles out of their way to avoid even the day-old smell of a human being. Bears don't seem particularly eager to encounter people, but when they do, they may react violently. Although it's an unusual behavior, polar bears, grizzlies, and even the smaller black bears have all occasionally killed people with the unmistakable motive of eating them. The hypotheses offered to account for human-eating range from the scent of menstrual blood to a shortage of blueberries. It's simpler, if more brutal, to recognize that a large predator may simply find us a serviceable source of protein.

It's possible to find books and articles only a couple of decades old that claim the mountain lion is no threat to human beings, that the few attacks that do take place are flukes. The past decade has rendered this position ridiculous.

For example: In January 1991, an eighteen-year-old man was jogging near Idaho Springs, Colorado, when a mountain lion attacked and killed him. In May 1992, in Kyuquot on Vancouver Island, a mountain lion took an eight-year-old boy from the edge of a schoolyard. In August 1996, a mountain lion attacked four people who were riding horses. A six-year-old boy seemed to be the main target of the attack. He survived; his mother died in his defense. From Chile to British Columbia, from California to Texas, encounters between people and mountain lions have been on the rise. Most often the encounters are harmless. Some are like mine — a person gets stalked but not hurt. One researcher told me mountain lions have the same sort of curiosity house cats do, and that's the best explanation of this pointless stalking. In other cases the meeting of human and feline is accidental or is initiated by the human. But an increasing number of encounters have resulted in violence, and some in death. In one recent attack, the mountain lion was rabid. But normally when mountain lions kill people, the motive is appetite. Most victims of mountain lion attacks are eaten.

Since the mountain lion ranges over so many nations, states, and provinces, it's hard to compile accurate figures on mountain lion attacks. One researcher who has tried to do so is Lee Fitzhugh of the University of California at Davis. Combining his own research with the earlier work of Paul Beier of Northern Arizona University, Fitzhugh came up with numbers that bear out the trend toward more human–mountain lion interactions. Between 1991 and the present, seven fatal attacks occurred. For the decade before that, only two fatal attacks were documented. The records compiled by Fitzhugh and Beier stretch from 1890 to the present. In that time, the average number of attacks per year has increased by a factor of fourteen.

This phenomenon is fairly mysterious, but some likely causes have emerged. We have, in effect, invaded mountain lion territory. We've built rural housing in areas that used to be utterly wild. The homeowner's desire for a little wilderness in his backyard sometimes succeeds too well. Outdoor activities like hiking have gained popularity steadily. Such forms of recreation can lead directly to human-lion encounters, and they can also familiarize the lions with the look and scent of humanity, decreasing any fear they may have of us. At the same time, environmental awareness and hunting regulations have made human-lion encounters less likely to be scary — for the lions.

Because mountain lions are territorial, their population increase means younger lions are pushed out into less desirable territories, such as those harboring human beings. Mountain lions have to learn which animals are suitable for prey. Some of this information comes from mothers and siblings, but much of it also comes from youthful experimentation. That's why younger animals are more dangerous to us. An older animal may disregard people as a food source simply because it never learned to see us that way. Younger mountain lions are more open to culinary experimentation.

One of Harris's neighbors, a rancher whose saddle horses had been attacked, obtained a permit to shoot the offending cats. He sat in a blind that evening and blew into a call that made a high-pitched chittering like that of an injured rabbit. He was astonished at his luck when a mountain lion stepped into clear view at dusk. She placed her broad feet with careful grace. Twice she stopped and looked into the distance, her black-tipped ears creeping forward slightly. He was even more astonished when a cub followed, and then three more. He fired on the mother lion, flipping her neatly with one shot. She screamed once, like a grieving woman, a scream that made the man shiver.

The cubs milled in confusion. The rancher shot again, dropping the first cub in convulsions. After missing his third shot, the man managed to kill two more cubs before the lone survivor slipped into the tall grass and escaped.

Harris has five children; the youngest is six, the oldest eighteen. Every time I've visited the ranch the children have been busy outside — burning trash, tending the tomato garden, even riding the range in search of stray cattle. I asked Mrs. Harris if she wasn't afraid to send the children out. She took me into the driveway, where her ten-year-old daughter was saddling two mares, a pinto and a sorrel. Mrs. Harris patted a rifle scabbard strapped to the pinto's saddle. The new leather of the scabbard stood out against the worn saddle. "We're sending them out with .22s, and they always ride in pairs."

Mountain lions are more likely to attack children than adults. They find taller animals more threatening. One conservation group has a Web page that advises anyone approached by a mountain lion to "try to appear larger by raising your hands." Running away and turning your back increase the danger. If you want to know how a mountain lion reacts to a person running — especially a small person — roll a ball in front of a pet cat. Besides looking large, the best ways to intimidate a mountain lion are growling, throwing rocks, and baring your teeth.

Once attacked, several people have successfully fought off mountain lions. In 1992, Morgan Boucke, a biologist with the California Department of Fish and Game, confronted a mountain lion in a backyard in Ojai. The lion charged. Boucke shot a couple of rounds in the air and then hit the lion on the head with a gun (another woman in the house also hit it with a lawn chair). Although the lion didn't give up the attack, Boucke survived uninjured until a warden arrived and the cat was shot.

Others have died in spite of fighting back. In a well-publicized case in El Dorado County, California, in 1994, a long-distance runner named Barbara Schoener was attacked on a trail. The lion knocked Schoener down an embankment to a creek, investigators determined later, where Schoener regained her footing and put up a fight. The lion eventually killed her.

It appears people who *don't* fight back *always* die. Playing dead only makes the hungry cat's job easier.

I asked Mrs. Harris whether she thought the lions were really a danger to the people in her community.

"It's just a matter of time," she said.

One late afternoon five months later, Harris went out to track a lion that took a neighbor's goat. The trail from the slain goat was fresh, and Harris's tracking dogs brought him into rocky terrain cut by ravines. He saw nothing, not even a pugmark.

"I knelt down and asked the Lord to show me where the mountain lion was," Harris said. Then he walked on and immediately spotted the animal. "He had to make a thirty-foot drop off a cliff or come right over me. I shot him while he was making up his mind."

It's legal in Colorado to shoot mountain lions out of season if they are destroying livestock or endangering people. But the law insists, among other provisions, that any stock-killing lion that is shot belongs to the state and must be reported. In a further conversation with the Lord, Harris determined that such a policy is wasteful. He phoned a taxidermist.

BLAINE HARDEN

The Dirt in the New Machine

FROM *The New York Times Magazine*

BEFORE YOU MAKE another call on that cell phone, take a moment, close your eyes, and reflect on all you've done for Mama Doudou, queen of the rain-forest whores.

Thanks to dollars that you and millions like you have spent on cell phones and Sony PlayStations, Mama Doudou had a knockout spring season in a mining camp called Kuwait, deep in central Africa. Kuwait — a name suggesting big money from below ground — was one of twenty illegal mines hacked in the past year out of the Okapi Faunal Reserve, a protected area in the Ituri rain forest of eastern Congo. The reserve is named after a reclusive, big-eared relative of the giraffe that is found only in Congo. Along with about four thousand okapi, the reserve is home to a rich assemblage of monkeys (thirteen species), an estimated ten thousand forest elephants, and about the same number of Mbuti people, often called pygmies, who live by hunting, gathering, and trading.

Mama Doudou, though, didn't mess with wildlife or pygmies. She sold overpriced bread in the mining camp and negotiated terms of endearment among three hundred miners and thirty-seven prostitutes. For a miner to secure the affections of a prostitute, he had to bring Mama Doudou some of the precious ore he was digging up in the reserve: a gritty, superheavy mud called coltan.

Coltan is abundant and relatively easy to find in eastern Congo. All a miner has to do is chop down great swaths of the forest, gouge SUV-size holes in streambeds with pick and shovel, and spend days up to his crotch in muck while sloshing water around in a plastic washtub until coltan settles to the bottom. (Coltan is three times heavier than iron, slightly lighter than gold.) If he is strong and relentless and the digging is good, a miner can produce a kilogram a day. Earlier this year, that was worth eighty dollars — a remarkable bounty in a region where most people live on twenty cents a day.

Coltan is the muck-caked counterpoint to the brainier-than-thou, environmentally friendly image of the high-tech economy. The wireless world would grind to a halt without it. Coltan, once it is refined in American and European factories, becomes tantalum, a metallic element that is a superb conductor of electricity, highly resistant to heat. Tantalum powder is a vital ingredient in the manufacture of capacitors, the electronic components that control the flow of current inside miniature circuit boards. Capacitors made of tantalum can be found inside almost every laptop, pager, personal digital assistant, and cell phone.

Mama Doudou, who is forty-five, is formally known as Doudou Wangonda, but she is called Mama because in the rain forest she is widely respected. She told me she doesn't understand what "rich white people" do with coltan. But she's exceptionally well versed in how much they pay for it. Late last year, exploding demand for tantalum powder created a temporary worldwide shortage, which contributed to Sony's difficulties in getting its new PlayStation 2 into American stores, as well as to a tenfold price increase on the world tantalum market. Mama Doudou abandoned her position as a traditional chief and joined thousands of people who walked into the Ituri forest hoping to get rich quick.

When the price of coltan was soaring, Mama Doudou made an absolute killing. First, she sold bread to miners at a scandalous price. She made as much as $800 worth of coltan for every $50 in cash that she spent on baking supplies. Then she used what she called her "natural leadership abilities" to win election as president of the camp prostitutes, most of whom were poorly educated, town bred women in their late teens. As president, Mama Doudou collected — and turned over to the owner of the mine — a variety of fees and fines related to the mating habits of miners and their women.

The normal arrangement in the camp was for a miner, after forking over a kilo of coltan to Mama Doudou, to pair off with one woman for the duration of their respective stays in the forest. The miner's "temporary wife" would cook his food, haul his water, and share his bed in a shack made of sticks and leaves. In return, he would give her enough coltan to keep her in cosmetics, clothes, and beer. If a miner decided that he wanted a prettier young woman to haul his water, he had to pay Mama Doudou another kilo of coltan.

"This is called the infringement fee," she explained.

Likewise, if a woman decided, as many did, to dump one miner in favor of another who happened to be a better producer of coltan, then she, too, had to pay Mama Doudou a kilo of coltan.

"This also is called the infringement fee," she said.

Frequent swapping of "temporary wives" in an equatorial forest where hygiene was problematic and condoms all but nonexistent led to an explosion of gonorrhea.

"There was too much sofisi," Mama Doudou said, using the Swahili word for the disease. Soon half the people in Kuwait had it. Antibiotics that could knock down gonorrhea were on sale in the camp for a tomato tin of coltan (worth about $27). They sold exceptionally well.

Mama Doudou's business ventures were part of a squalid encounter between the global high-tech economy and one of the world's most thoroughly ruined countries.

Congo — always too well endowed with natural resources and too weakly governed for its own good — is a nation in name only. The Democratic Republic of the Congo is in the late stages of a political malady that students of modern Africa call "failed state syndrome." Roads, schools, and medical clinics barely exist. Malnutrition and poverty have brought back diseases, like sleeping sickness, that had been under control. The World Health Organization recently estimated that the monthly toll of "avoidable deaths" in Congo was 72,800.

The eastern half of the country, with about 20 million residents, has no real government and no laws except the ever-changing rules imposed by invading armies from Rwanda and Uganda and roving bands of well-armed predators.

A scalding report that was presented this spring to the United Nations Security Council said that coltan perpetuates Congo's civil war. The report, based on a six-month investigation by an expert panel, said the war "has become mainly about access, control, and trade" of minerals, the most important being coltan. The one thing that unites the warring parties, according to the report, is a keen interest in making money off coltan.

"Because of its lucrative nature," the report said, the war "has created a 'win-win' situation for all belligerents. Adversaries and enemies are at times partners in business, get weapons from the same dealers, and use the same intermediaries. Business has superseded security concerns."

Environmental groups have added emotional fuel to the accusations in the UN report by cataloguing the devastation that the coltan trade has brought to Congo's wildlife. About ten thousand miners and traders have overrun Kahuzi-Biega National Park, according to a report released in May by a coalition of environmental groups. Before the civil war, the park was home to about eight thousand eastern lowland gorillas. That number may have since been reduced to fewer than one thousand, the report estimated, because miners and others in the forest are far from food supplies and must rely on bush meat. Apes are killed for food or killed in traps set for other animals. If something is not done to stop mining and poaching, the report said, the eastern lowland gorilla "will become the first great ape to be driven to extinction — a victim of war, human greed, and high technology."

While coltan extraction has taken advantage of Congo's ruin, it did not cause it. That has taken more than 110 years of misrule, during which Congo has attracted a string of shady suitors.

The most malign of the courtships began in the late nineteenth century, when agents of King Leopold II of the Belgians started stripping central Africa of ivory and rubber. To enforce production quotas on the locals, Leopold's agents chopped off their hands, noses, and ears. Before the king was forced to trade away the hugely profitable colony in 1908, an estimated five million to eight million Congolese were killed.

In the 1960s, the Americans waded in. To fight Communism and secure access to cobalt and copper, the Central Intelligence Agency helped bring about the assassination of Congo's first democratically elected prime minister, Patrice Lumumba. That was followed by three decades of White House coddling of his successor, Mobutu Sese Seku, Africa's most famous billionaire dictator, who set a poisoned table for the chaos that followed his eventual overthrow in 1997.

Since then, Congo has been locked in a sprawling and numbingly complicated civil war that by some estimates has become the deadliest conflict in the history of independent Africa. The war has caused the deaths of 2.5 million people over the past two and a half years in eastern Congo alone, according to a recent report by the International Rescue Committee, a New York–based aid agency, which described the emergency in Congo as "perhaps worse than any to unfold in Africa in recent decades."

The coltan story seemed clear when I flew to Congo early this summer. Globalization was causing havoc in a desperate country. For the sake of our electronic toys, guerrillas were getting rich, gorillas were getting slaughtered, and the local people were getting paid next to nothing to ruin their country's environment. Traveling inside Congo, however, I found clarity on the question of coltan to be as scarce as paved roads, functioning schools, or sober soldiers.

What muddied up the story, first of all, was the curiously egalitarian quality of coltan mining. Just about anyone with a shovel and a strong back can dig it up. It's easier to find and more plentiful than diamonds, which have created their own blood frenzy in Africa. It has injected hundreds of millions of dollars into an economy that had virtually ceased to function. True, much of that money has been creamed off by warlords and profiteers, and very little of it has been redistributed in social services. Some, though, has filtered down to miners, middlemen, and commerçants.

To discover the importance of coltan's trickle-down effect (and to meet Mama Doudou), I first had to take a spine-mashing, thirteen-hour ride on the back of a Yamaha trail bike over a mud track that used to be the main east-west highway across northern Congo.

The road I traveled is all but impassable to motorized vehicles, excepting a trail bike driven by someone who knows how to negotiate mammoth mudholes, many of which are deeper and longer than a New York City garbage truck, as well as when to bribe drunken rogue soldiers and when to run from them.

Riding through the reserve, I was menaced by a Congolese soldier, a member of the Front for the Liberation of Congo, a poorly disciplined rebel group supported by the Ugandan Army. He demanded my boots, explaining that he didn't have boots. He demanded money, explaining that he had none. He pointed his AK-47 at my stomach. Gunning the Yamaha, my driver sped away before the soldier, who was stumble-down drunk, could react. This encounter, my driver later explained, was normal.

In Epulu, a village that is the administrative center of the Okapi reserve, I spent an afternoon with a coltan miner named Munako Bangazuna, a quiet, wary man who stood only four feet six inches tall. Early this year, Bangazuna enjoyed what he called "my richest period." In a mine inside the Okapi reserve, he dug about a kilo of coltan a day, he said. Working seven days a week, he made more than $2,000 a month for two months in a row — a fortune in the forest.

He is twenty-six and married with children. Mining allowed him to provide his family with food and consumer goods he never dreamed he'd be able to afford. Besides food, he bought a bicycle, a radio, a foam mattress, cooking pots, dishes, and clothes for himself, his wife, and his kids.

Bangazuna does not claim to have spent his money wisely. The mining camp where he lived was called Boma Libala, a phrase that means "kill the marriage." It was the largest and, by reputation, nastiest mining camp in the reserve, with three thousand miners and several hundred prostitutes.

"I lost a lot of my money on prostitution and also on Primus," said Bangazuna, referring to a brand of Congolese beer. His wife cooked for him in Boma Libala, he said, during the time he was drinking lots of beer and spending most of his money on prostitutes. "I was lucky," he said. "She did not divorce me."

Bangazuna was hardly alone in his bad behavior. One coltan moment that particularly nauseates authorities occurred this spring in Epulu, when a drunken miner and a seminaked prostitute fornicated in broad daylight on the lawn of the primary school, in front of village children.

Like several miners I interviewed in the reserve, a territory controlled by the Ugandan military and its rebel allies, Bangazuna was compelled to give up a slice of his coltan diggings to an extortion racket run by Ugandan soldiers.

"In the morning, when you get up, the Ugandans hand you a pack of cigarettes, and they give you two bottles of beer," said Bangazuna, explaining his daily routine. "In the evening, when you finish digging, you have to pay them back with coltan. It was very expensive. One bottle of beer cost me two spoons of coltan" —

about eight dollars — and cigarettes were one spoon. If you refuse to pay or if you don't have coltan, they beat you and threaten to shoot you."

When I talked to Bangazuna, he was broke. He had spent all the money he'd earned digging coltan. He also happened to be under arrest. Game wardens (whose expenses are paid partly by donations from several American zoos) had caught him digging coltan in the reserve after he'd been warned not to do so.

When the wardens let him go, Bangazuna confided, he planned to dig more coltan.

This spring, the price of coltan crashed, falling from $80 a kilo in March to $8 in June. As cell phone sales slumped and the Nasdaq shrank, demand for coltan from companies like Nokia, Ericsson, and Motorola fell precipitously. Suddenly, a Congolese coltan miner had to dig coltan all day simply to afford to eat in a mining camp, and he had to dig for three or four days to find enough coltan to pay for the drugs that would clear up a case of sofisi.

At the Kuwait mine, everything came unglued. The prostitutes, then the merchants, then the miners and, finally, Mama Doudou herself abandoned the mine and walked out of the rain forest.

A few days before game wardens burned it to the ground, I visited another mine in the Okapi reserve. It was more accessible than Kuwait and, unlike the infamous but shut-down Boma Libala camp, where Bangazuna had squandered so much of his money, it still had a few working miners.

To get to Tuko-Tu camp, I walked for about two hours on a well-trod trail beneath a high canopy of trees. They shrouded the rain forest in permanent shadow. It rained hard as I walked, and the dark, soggy forest was threaded with filigrees of mist.

For all its gloom, the forest was about as primeval as a rest stop on the New Jersey Turnpike. Every half hour or so, two or three giggly women, seemingly dressed for a party in glistening lipstick and gaudy dresses, smelling of strong perfume, would materialize out of the dank greenery. Their bare feet were caked with mud; they carried their shoes. They were prostitutes from Tuko-Tu, and they were walking out to buy bread and beer in a village on the main road.

On both sides of the footpath, a towering mainstay of the forest was dying for the sake of coltan. The eko, one of three giant trees that form the rain forest's canopy, has a durable, waterproof bark. Miners strip off a long girdle of the bark to make a trough into which they shovel coltan-bearing mud, which is then flushed with water. Stripping bark has killed thousands of eko trees in the reserve, which greatly upsets the local pygmies. They rely on the tree, whose flowers attract bees, as depots for gathering honey.

The mining camp squatted in a clearing hacked out of the forest. It was a jumble of stick huts with roofs made from forest leaves. Most of the huts were empty. The camp was down to 57 residents, from a high of 320 when coltan was at $80 a kilo. Out in front of the huts, a few toddlers stood stoically in the mud as bored young mothers picked lice from their hair. Ugandan soldiers used to come here, I was told, to force miners to buy beer and cigarettes. But they had stopped coming in May, when the price of coltan began to fizzle. There were still miners, of course, but they were out digging, deeper in the forest.

So I walked on, following a streambed. After about an hour and a half of walking, the streambed suddenly disappeared. A bombing range took its place — or what looked like a bombing range. Craters, hundreds of them, many ten feet deep, all of them partly filled with muck, marched on for miles through the forest — until they ended in a cluster of about twenty-five miners. With shovels, picks, and plastic washtubs, they were creating more craters in the streambed.

Jean Pierre Asikima, forty-three, was among them. He gave up digging gold two months earlier, he said, to come into the forest in search of coltan. But the digging was poor, the price was low, and he said he was losing weight in the forest.

"I have come too late, and soon I will quit," he said, standing in a crater he had dug, waist-deep in muddy water that was the color of chicken gravy.

Still, Asikima worked with a fury. With mud and gravel from his crater, he built a seven-foot-high mound. Then he shoveled and scraped the mud into an eko-bark trough, while another miner, a sixteen-year-old who said his name was Dragon, went to work with a blue plastic washtub, pouring several hundred tubs full of water through the makeshift sluice.

After about an hour, a glittering black stain had gathered at the foot of the trough. It was an ounce or so of coltan, the fruit of five hours of digging and washing. Asikima said it was not enough to buy a tin of rice for dinner. As he began to dig another crater, I asked him about his life in the forest. He said he despised it.

"If I had another job," he said, "I would not come here. But there are no other jobs. When this mine closes, I will go and find another one."

He is not alone. Although the price has crashed, many coltan-bearing regions of eastern Congo remain thick with miners, commerçants, and prostitutes. In a country with a twenty-cents-a-day living standard, the chance of earning a few dollars from coltan is still a powerful enough reason to live in the bush and shovel muck, sell bread, and risk sofisi.

To halt war profiteering and the destruction of wildlife, the report to the UN Security Council called for an embargo on the export of coltan and other natural resources from Uganda and Rwanda.

Although the embargo has yet to be imposed by the Security Council, European and American companies that profit from the coltan trade have been scurrying to avoid bad publicity. Pictures of dead gorillas and of environmental ruin in Congo's national parks have been particularly effective in triggering alarm among companies that pride themselves on their environmental images.

Sabena, the Belgian airline named in the report for hauling Congolese coltan to Europe, announced in June that it will no longer carry the ore. Nokia and Motorola are among several major mobile-phone makers that have publicly demanded that their suppliers stop using ore mined illegally in Congo. And so it has gone down the supply chain. The world's largest maker of tantalum capacitors, Kemet, in Greenville, South Carolina, has asked its suppliers to certify that ore does not come from Congo or bordering countries, including Uganda and Rwanda. Cabot Corporation, a Boston-based company that is the world's second-largest processor of tantalum powder, announced this spring that it "deplores all unlawful and immoral activities" connected with coltan mined in Congo and declared that it will not buy any ore from that part of the world.

The high moral ground that companies have been quick to stake out has the added attraction of being profit-neutral, at least for the moment. The tantalum market is glutted because of declining demand in the slumping technology sector. As important, there has been a sharp increase in production by a giant Australian mining company, called Sons of Gwalia, which now produces half the world's supply.

In the immediate future, it looks as though less and less of the world's tantalum will come from Congo's coltan mines. And as the UN sees it, this will be a very good thing. Its report concluded, "The only loser in this huge business venture is the Congolese people."

But inside what's left of Congo, a wide array of influential people (who are not combatants in the war and are not getting rich from coltan) make a persuasive case that these demands are naive and could well produce disastrous consequences. They argue that in a collapsed state where the likelihood of constructive Western intervention is next to nil, there simply are no easy fixes.

"For local people who are trying to make a bit of money out of coltan, how can an embargo possibly help?" asked Aloys Tegera, who directs the Pole Institute, a nongovernmental social-research institute in Goma, in eastern Congo. Tegera is well aware of coltan's destructive side: he is the lead author of a study on the severe social impact of coltan mining, which describes how teachers have been lured to the mines from the country's few functioning classrooms and explains why teenage girls have turned to prostitution.

"Coltan fuels the war; nobody can deny that," said Tegera. "That is why we maybe will never get peace. But civilians, especially those who are organized, also are getting some money from this."

He and many others find it more than slightly insulting that in a country where millions are hungry and coltan is helping to feed some of them, a de facto embargo is gathering steam among high-tech companies apparently worried less about human beings than about the public-relations downside of dead gorillas. And, like many other Congolese, he declines to become morally riled up about foreign domination.

"Of course, the Rwandans are pillaging us," he said. "But they are not the first to do it and they are no worse than the others. King Leopold did it. The Belgians did it. Mobutu and the Americans did it. The most sorrowful thing I have to live with is that we are incapable of coming up with an elite that can run things with Congolese interests in mind."

Terese Hart, an American botanist who helped create the Okapi Faunal Reserve and has worked there since the early 1980s, supports neither an embargo on coltan nor a quick pullout of Ugandan forces from northeast Congo.

"The world wants to intervene from a distance and pull the strings on the puppet," said Hart, who works for the Wildlife Conservation Society. "The problem is that the strings are not connected to anything. When outsiders struggle to find solutions for Congo, they often assume there is some kind of government. There is no government. There is nothing."

As for coltan mining, Hart said it is silly for the outside world to try to squeeze one of the few ways for poor people to make a bit of money.

"Outside the reserve, I think that coltan mining is the lesser evil of the types of exploitation that occur when there is no government," Hart said. "I prefer mining to logging. Cutting timber in the rain forest is part of an irreversible ecological process. I don't think coltan mining does as much permanent damage. The miner will not get much, but at least he will continue to live."

Among the Congolese I spoke to about coltan, the consensus was that they could not risk the simple solutions that outsiders had prescribed. Struggling to survive in a failed state, they saw no straightforward answers, no moral high ground. For them, the only thing worse than mining coltan is not mining it.

What progress there is in eastern Congo tends to be slow, small-scale, and subject to sudden reversal.

The World Health Organization has succeeded in working with rebel groups to vaccinate most children for polio, but it says seven of ten children have not received any other vaccines in the past decade. Western donors are distributing some medicines, seeds, and tools, but three-quarters of the population still has no access to basic health care.

When progress is being made, it often involves the mixed blessing of coltan. In eastern Congo, two mining entrepreneurs, Edouard Mwangachuchu, a Congolese Tutsi, and his American partner, Robert Sussman, a physician from Baltimore, are struggling to build a legitimate business in an illegitimate state.

They run a company that even their competitors say treats miners fairly. It supplies shovels and picks to about a thousand men who operate as independent contractors in mines located far from national parks, protected forests, and endangered gorillas.

The land belongs to Mwangachuchu, whose herds were slaughtered in 1995, as the Mobutu era was sputtering to an end. Desperate to shore up popular support, Mobutu encouraged Congolese in the east to attack the ethnic Tutsi minority. A mob pulled Mwangachuchu, then a financial adviser to the provincial government in Goma, out of his Suzuki jeep on his way to work. They choked him with his necktie, ripped off his clothes, and dumped him at the Rwandan border. Crowds later stoned and shot at his house.

Mwangachuchu, his wife, and their six children were granted political asylum in the United States in 1996, and they rented a house in Laurel, Maryland. Two years later, homesick and bored with his job at a Carvel ice cream plant, Mwangachuchu returned home for a visit.

Civil war was raging. His cattle farm had been destroyed, his herds gone, his buildings burned. But he still owned the land, which he had long known was rich in coltan. In 1999, a year after his first trip home, he heard that there was money to be made mining it. All he needed was a partner, someone with a bit of money.

Robert Sussman, fifty-five, sold his medical practice in Baltimore in the early 1990s. Comfortably well off, he began a second career as a mining-camp doctor in remote countries, including Myanmar and Congo. Intrigued by mining, he began thinking of going into the business himself. He met Mwangachuchu in a Goma hotel in 1998, and they became partners the following year, as the price of coltan began to go up.

Sussman and Mwangachuchu say they are investing in Congo for the long term. They believe they can operate profitably despite the recent slump in coltan prices and despite the fact that their mines are still periodically fought over by roving bands of armed men. The partners say they have laid the foundation for a solid business.

"We are proud of what we are doing in Congo," Sussman told me. "We want the world to understand that if it's done right, coltan can be good for this country."

Sussman and Mwangachuchu, of course, are also in it for the money. High coltan prices last year gave them an unexpected windfall. Sussman said they sold twenty-two metric tons of coltan, which earned them about $7.5 million — before they paid their many bills.

Since then, they have bought about twenty-five more tons of coltan from miners in the field — ore that they have not been able to sell.

Last year, Sussman and Mwangachuchu shipped their ore to Europe on Sabena airlines. That airline now refuses their business, and they are scrambling to find another shipper. They fear that a corporate embargo could cripple their business and idle miners who have come to depend on them.

"We don't understand why they are doing this," Mwangachuchu told me. "The Congolese have a right to make business in their own country."

ROBERT M. HAZEN

Life's Rocky Start

FROM *Scientific American*

NO ONE KNOWS how life arose on the desolate young earth, but one thing is certain: life's origin was a chemical event. Once the earth formed 4.5 billion years ago, asteroid impacts periodically shattered and sterilized the planet's surface for another half a billion years. And yet, within a few hundred million years of that hellish age, microscopic life appeared in abundance. Sometime in the interim, the first living entity must have been crafted from air, water, and rock.

Of those three raw materials, the atmosphere and oceans have long enjoyed the starring roles in origins-of-life scenarios. But rocks, and the minerals of which they are made, have been called on only as bit players or simply as props. Scientists are now realizing that such limited casting is a mistake. Indeed, a recent flurry of fascinating experiments is revealing that minerals play a crucial part in the basic chemical reactions from which life must have arisen.

The first act of life's origin story must have introduced collections of carbon-based molecules that could make copies of themselves. Achieving even this nascent step in evolution entailed a sequence of chemical transformations, each of which added a level of structure and complexity to a group of organic molecules. The most abundant carbon-based compounds available on the ancient earth were gases with only one atom of carbon per molecule, namely, carbon dioxide, carbon monoxide, and methane. But the essential building blocks of living organisms — energy-rich sugars, membrane-forming lipids, and complex amino acids — may include more than a dozen carbon atoms per molecule. Many of these molecules, in turn, must bond together to form chainlike polymers and other molecular arrays in order to accomplish life's chemical tasks. Linking small molecules into these complex, extended structures must have been especially difficult in the harsh conditions of the early earth, where intense ultraviolet radiation tended to break down clusters of molecules as quickly as they could form.

Carbon-based molecules needed protection and assistance to enact this drama. It turns out that minerals could have served at least five significant functions, from passive props to active players, in life-inducing chemical reactions. Tiny compartments in mineral structures can shelter simple molecules, while mineral surfaces can provide the scaffolding on which those molecules assemble and grow. Beyond these sheltering and supportive functions, crystal faces of certain minerals can actively select particular molecules resembling those that were destined to become biologically important. The metallic ions in other minerals can jump-start meaningful reactions like those that must have converted simple molecules into self-replicating entities. Most surprising, perhaps, are the recent indications that elements of dissolved minerals can be incorporated into biological molecules. In other words, minerals may not have merely helped biological molecules come together, they might have become part of life itself.

For the better part of a century, following the 1859 publication of Charles Darwin's *On the Origin of Species,* a parade of scientists speculated on life's chemical origins. Some even had the foresight to mention rocks and minerals in their inventive scenarios. But experimental evidence only sporadically buttressed these speculations.

One of the most famous experiments took place at the University of Chicago in 1953. That year chemist Harold C. Urey's precocious graduate student Stanley L. Miller attempted to mimic the earth's primitive oceans and atmosphere in a bottle. Miller enclosed methane, ammonia, and other gases thought to be components of the early atmosphere in a glass flask partially filled with water. When he subjected the gas to electric sparks to imitate a prehistoric lightning storm, the clear water turned pink and then brown as it became enriched with amino acids and other essential organic molecules. With this simple yet elegant procedure, Miller transformed origins-of-life research from a speculative philosophical game to an exacting experimental science. The popular press sensationalized the findings by suggesting that synthetic bugs might soon be crawling out of test tubes. The scientific community was more restrained, but many workers sensed that the major obstacle to creating life in the laboratory had been solved.

It did not take long to disabuse researchers of that notion. Miller may have discovered a way to make many of life's building blocks out of the earth's early supply of water and gas, but he had not discovered how or where these simple units would have linked into the complex molecular structures — such as proteins and DNA — that are intrinsic to life.

To answer that riddle, Miller and other origins scientists began proposing rocks as props. They speculated that organic molecules, floating in seawater, might have splashed into tidal pools along rocky coastlines. These molecules would have become increasingly concentrated through repeated cycles of evaporation, like soup thickening in a heated pot.

In recent years, however, researchers have envisioned that life's ingredients might have accumulated in much smaller containers. Some rocks, like gray volcanic pumice, are laced with air pockets created when gases expanded inside the rock while it was still molten. Many common minerals, such as feldspar, develop microscopic pits during weathering. Each tiny chamber in each rock on the early earth could have housed a separate experiment in molecular self organization. Given enough time and enough chambers, serendipity might have produced a combination of molecules that would eventually deserve to be called "living."

Underlying much of this speculation was the sense that life was so fragile that it depended on rocks for survival. But in 1977 a startling discovery challenged conventional wisdom about life's fragility and, perhaps, its origins. Until then, most scientists had assumed that life spawned at or near the benign ocean surface as a result of chemistry powered by sunlight. That view began to change when deep-ocean explorers first encountered diverse ecosystems thriving at the superheated mouths of volcanic vents on the sea floor. These extreme environments manage to support elaborate communities of living creatures in isolation from the sun. In these dark realms, much of the energy that organisms need comes not from light but from the earth's internal heat. With this knowledge in mind, a few investigators began to wonder whether organic reactions relevant to the origins of life might occur in the intense heat and pressure of these so-called hydrothermal vents.

Miller and his colleagues have objected to the hydrothermal origins hypothesis in part because amino acids decompose rapidly when they are heated. This objection, it turns out, may be applicable only when key minerals are left out of the equation. The idea that minerals might have sheltered the ingredients of life received a boost from recent experiments conducted at my home base, the Carnegie Institution of Washington's Geophysical Laboratory. As a postdoctoral researcher at Carnegie, my colleague Jay A. Brandes (now at the University of Texas Marine Sciences Institute in Port Aransas) proposed that minerals help delicate amino acids remain intact. In 1998 we conducted experiments in which the amino acid leucine broke down within a matter of hours in pressurized water at 200 degrees Celsius. But when Brandes added to the mix an iron sulfide mineral of the type found in and around hydrothermal vents, the amino acid stayed intact for much longer, giving leucine more time to react with other molecules.

Even if the right raw materials were contained in a protected place — whether it was a tidal pool, a microscopic pit in a mineral surface, or somewhere inside the plumbing of a sea floor vent — the individual molecules would still be suspended in water. These stray molecules needed a support structure — some kind of scaffolding — where they could cling and react with one another.

One easy way to assemble molecules from a dilute solution is to concentrate them on a flat surface. Errant molecules might have been drawn to the calm surface of a tidal pool or perhaps to a primitive "oil slick" of compounds trapped at the water's surface. But such environments would have posed a potentially fatal hazard to delicate molecules. Harsh lightning storms and ultraviolet radiation accosted the young earth in doses many times greater than they do today. Such conditions would have quickly broken the bonds of complex chains of molecules.

Origins scientists with a penchant for geology have long recognized that minerals might provide attractive alternative surfaces where important molecules could assemble. Like the container idea, this notion was born half a century ago. At that time, a few scientists had begun to suspect that clays have special abilities to attract organic molecules. These ubiquitous minerals feel slick when wet because their atoms form flat, smooth layers. The surfaces of these layers frequently carry an electric charge, which might be able to attract organic molecules and hold them in place. Experiments later confirmed these speculations. In the late 1970s an Israeli research group demonstrated that amino acids can concentrate on clay surfaces and then link up into short chains that resemble biological proteins. These chemical reactions occurred when the investigators evaporated a water-based solution containing amino acids from a vessel containing clays — a situation not unlike the evaporation of a shallow pond or tidal pool with a muddy bottom.

More recently, separate research teams led by James P. Ferris of the Rensselaer Polytechnic Institute and by Gustaf Arrhenius of the Scripps Institution of Oceanography demonstrated that clays and other layered minerals can attract and assemble a variety of organic molecules. In a tour-de-force series of experiments during the past decade, the team at Rensselaer found that clays can act as scaffolds for the building blocks of RNA, the molecule in living organisms that translates genetic instructions into proteins.

Once organic molecules had attached themselves to a mineral scaffold, various types of complex molecules could have been forged. But only a chosen few were eventually incorporated into living cells. That means that some kind of template must have selected the primitive molecules that would become biologically important. Recent experiments show, once again, that minerals may have played a central role in this task.

Perhaps the most mysterious episode of selection left all living organisms with a strange predominance of one type of amino acid. Like many organic molecules, amino acids come in two forms. Each version comprises the same types of atoms, but the two molecules are constructed as mirror images of each other. The phenomenon is called chirality, but for simplicity's sake scientists refer to the two versions as "left-handed" (or "L") and "right-handed" (or "D"). Organic synthesis experiments like Miller's invariably produce 50–50 mixtures of L and D molecules, but the excess of left-handed amino acids in living organisms is nearly 100 percent.

Researchers have proposed a dozen theories — from the mundane to the exotic — to account for this bizarre occurrence. Some astrophysicists have argued that the earth might have formed with an excess of L amino acids — a consequence of processes that took place in the cloud of dust and gas that became the solar system. The main problem with this theory is that in most situations such processes yield only the slightest excess — less than one percent — of L or D molecules.

Alternatively, the world might have started with a 50–50 mixture of L and D amino acids, and then some important feature of the physical environment selected one version over the other. To me, the most obvious candidates for this specialized physical environment are crystal faces whose surface structures are mirror images of each other. Last spring I narrowed in on calcite, the common mineral that forms limestone and marble, in part because it often displays magnificent pairs of mirror-image faces. The chemical structure of calcite in many mollusk shells bonds strongly to amino acids. Knowing this, I began to suspect that calcite surfaces may feature chemical bonding sites that are ideally suited to only one type of amino acid or the other. With the help of my Carnegie colleague Timothy Filley (now at Purdue University) and Glenn Goodfriend of George Washington University, I ran more than a hundred tests of this hypothesis.

Our experiments were simple in concept, although they required meticulous clean-room procedures to avoid contamination by the amino acids that exist everywhere in the environment. We immersed a well-formed, fist-size crystal of calcite into a 50–50 solution of aspartic acid, a common amino acid. After twenty-four hours we removed the crystal from this solution, washed it in water, and carefully collected all the molecules that had adhered to specific crystal faces. In one experiment after another we observed that calcite's "left-handed" faces selected L amino acids, and vice versa, with excesses approaching 40 percent in some cases.

Curiously, calcite faces with finely terraced surfaces displayed the greatest selectivity. This outcome led us to speculate that these terraced edges might force the L and D amino acids to line up in neat rows on their respective faces. Under the right environmental conditions, these organized rows of amino acids might chemically join to form protein-like molecules — some made entirely of L amino acids, others entirely of D. If protein formation can indeed occur, this result becomes even more exciting, because recent experiments by other investigators indicate that some proteins can self-replicate. In the earth's early history, perhaps a self-replicating protein formed on the face of a calcite crystal.

Left- and right-handed crystal faces occur in roughly equal numbers, so chiral selection of L amino acids probably did not happen everywhere in the world at once. Our results and predictions instead suggest that the first successful set of self-replicating molecules — the precursor to all the varied life-forms on the earth today — arose at a specific time and place. It was purely chance that the successful molecule developed on a crystal face that preferentially selected left-handed amino acids over their right-handed counterparts.

Minerals undoubtedly could have acted as containers, scaffolds, and templates that helped to select and organize the molecular menagerie of the primitive earth. But many of us in origins research suspect that minerals played much more active roles, catalyzing key synthesis steps that boosted the earth's early inventory of complex biological molecules.

Experiments led by Carnegie researcher Brandes in 1997 illustrate this idea. Biological reactions require nitrogen in the form of ammonia, but the only common nitrogen compound thought to have been available on the primitive earth is nitrogen gas. Perhaps, Brandes thought, the environment at hydrothermal vents mimics an industrial process in which ammonia is synthesized by passing nitrogen and hydrogen over a hot metallic surface. Sure enough, when we subjected hydrogen, nitrogen, and the iron oxide mineral magnetite to the pressures and temperatures characteristic of a sea floor vent, the mineral catalyzed the synthesis of ammonia.

The idea that minerals may have triggered life's first crucial steps has emerged most forcefully from the landmark theory of chemist Günter Wächtershäuser, a German patent lawyer with a deep interest in life's origins. In 1988 Wächtershäuser advanced a sweeping theory of organic evolution in which minerals — mostly iron and nickel sulfides that abound at deep-sea hydrothermal vents — could have served as the template, the catalyst, and the energy source that drove the formation of biological molecules. Indeed, he has argued that primitive living entities were molecular coatings that adhered to the positively charged surfaces of pyrite, a mineral composed of iron and sulfur. These entities, he further suggests, obtained energy from the chemical reactions that produce pyrite. This hypothesis makes sense in part because some metabolic enzymes — the molecules that help living cells process energy — have at their core a cluster of metal and sulfur atoms.

For much of the past three years, Wächtershäuser's provocative theory has influenced our experiments at Carnegie. Our team, including geochemist George Cody and petrologist Hatten S. Yoder, has focused on the possibility that metabolism can proceed without enzymes in the presence of minerals — especially oxides and sulfides. Our simple strategy, much in the spirit of Miller's famous experiment, has been to subject ingredients known to be available on the young earth — water, carbon dioxide, and minerals — to a controlled environment. In our case, we try to replicate the bone-crushing pressures and scalding temperatures typical of a deep-sea hydrothermal vent. Most of our experiments test the interactions among ingredients enclosed in welded gold capsules, which are roughly the size of a daily vitamin pill. We place as many as six capsules into Yoder's "bomb" — a massive steel pressure chamber that squeezes the tiny capsules to pressures approaching 2,000 atmospheres and heats them to about 250 degrees C.

One of our primary goals in these organic-synthesis experiments — and one of life's fundamental chemical reactions — is carbon fixation, the process of producing molecules with an increasing number of carbon atoms in their chemical structure. Such reactions follow two different paths depending on the mineral we use. We find that many common minerals, including most oxides and sulfides of iron, copper, and zinc, promote carbon addition by a routine industrial process known as Fischer-Tropsch (F-T) synthesis.

This process can build chainlike organic molecules from carbon monoxide and hydrogen. First, carbon monoxide and hydrogen react to form methane, which has one carbon atom. Adding more carbon monoxide and hydrogen to the methane produces ethane, a two-carbon molecule, and then the reaction repeats itself, adding a carbon atom each time. In the chemical industry, researchers have harnessed this reaction to manufacture molecules with virtually any desired number of carbon atoms. Our first organic-synthesis experiments in 1996, and much more extensive research by Thomas McCollom of the Woods Hole Oceanographic Institution, demonstrate that F-T reactions can build molecules with thirty or more carbon atoms under some hydrothermal-vent conditions in less than a day. If this process manufactures large organic molecules from simple inorganic chemicals throughout the earth's hydrothermal zones today, then it very likely did so in the planet's pre-biological past.

When we conduct experiments using nickel or cobalt sulfides, we see that carbon addition occurs primarily by carbonylation — the insertion of a carbon and oxygen molecule, or carbonyl group. Carbonyl groups readily attach themselves to nickel or cobalt atoms, but not so strongly that they cannot link to other molecules and jump ship to form larger molecules. In one series of experiments, we observed the lengthening of the nine-carbon molecule nonyl thiol to form ten-carbon decanoic acid, a compound similar to the acids that drive metabolic reactions in living cells. What is more, all the reactants in this experiment — a thiol, carbon monoxide, and water — are readily available near sulfide-rich hydrothermal vents. By repeating these simple kinds of reactions — adding a carbonyl group here or a hydroxide group there — we can synthesize a rich variety of complex organic molecules.

Our 1,500 hydrothermal organic synthesis experiments at Carnegie have done more than supplement the catalogue of interesting molecules that must have been produced on the early earth. These efforts reveal another, more complex behavior of minerals that may have significant consequences for the chemistry of life. Most previous origins-of-life studies have treated minerals as solid and unchanging — stable platforms where organic molecules could assemble. But we are finding that in the presence of hot water at high pressure, minerals start to dissolve. In the process, the liberated atoms and molecules from the minerals can become crucial reactants in the primordial soup.

Our first discovery of minerals as reactants was an unexpected result of our recent catalysis experiments led by Cody. As expected, carbonylation reactions produced ten-carbon decanoic acid from a mixture of simple molecules inside our gold capsules. But significant quantities of elemental sulfur, organic sulfides, methyl thiol, and other sulfur compounds appeared as well. The sulfur in all these products must have been liberated from the iron sulfide mineral.

Even more striking was the liberation of iron, which brilliantly colored the water-based solutions inside the capsules. As the mineral dissolved, the iron formed bright red and orange organometallic complexes in which iron atoms are surrounded by various organic molecules. We are now investigating the extent to which these potentially reactive complexes might act as enzymes that promote the synthesis of molecular structures.

The role of minerals as essential chemical ingredients of life is not entirely unexpected. Hydrothermal fluids are well known to dissolve and concentrate mineral matter. At deep-sea vents, spectacular pillars of sulfide grow dozens of feet tall as plumes of hot, mineral-laden water rise from below the sea floor, contact the frigid water of the deep ocean, and deposit new layers of minerals on the growing pillar. But the role of these dissolved minerals has not yet figured significantly in origins scenarios. Whatever their behavior, dissolved minerals seem to make the story of life's emergence much more interesting.

When we look beyond the specifics of pre-biological chemistry, it is clear that the origin of life was far too complex to imagine as a single event. Rather we must work from the assumption that it was a gradual sequence of more modest events, each of which added a degree of order and complexity to the world of pre-biological molecules. The first step must have been the synthesis of the basic building blocks. Half a century of research reveals that the molecules of life were manufactured in abundance — in the nebula that formed our solar system, at the ocean's surface, and near hydrothermal vents. The ancient earth suffered an embarrassment of riches — a far greater diversity of molecules than life could possibly employ.

Minerals helped to impose order on this chaos. First by confining and concentrating molecules, then by selecting and arranging those molecules, minerals may have jump-started the first self-replicating molecular systems. Such a system would not have constituted life as we know it, but it could have, for the first time, displayed a key property of life. In this scenario, a self-replicating molecular system began to use up the resources of its environment.

As mutations led to slightly different variants, competition for limited resources initiated and drove the process of molecular natural selection. Self-replicating molecular systems began to evolve, inevitably becoming more efficient and more complex.

A long-term objective for our work at the Carnegie Institution is to demonstrate simple chemical steps that could lead to a self-replicating system — perhaps one related to the metabolic cycles common to all living cells. Scientists are far from creating life in the laboratory, and it may never be possible to prove exactly what chemical transformations gave rise to life on earth. What we can say for sure is that minerals played a much more complex and integral part in the origin of life than most scientists ever suspected. By being willing to cast minerals in starring roles in experiments that address life's beginnings, researchers may come closer to answering one of science's oldest questions.

SARAH BLAFFER HRDY

Mothers and Others

FROM *Natural History*

MOTHER APES — chimpanzees, gorillas, orangutans, humans — dote on their babies. And why not? They give birth to an infant after a long gestation and, in most cases, suckle it for years. With humans, however, the job of providing for a juvenile goes on and on. Unlike all other ape babies, ours mature slowly and reach independence late. A mother in a foraging society may give birth every four years or so, and her first few children remain dependent long after each new baby arrives; among nomadic foragers, grown-ups may provide food to children for eighteen or more years. To come up with the 10–13 million calories that anthropologists such as Hillard Kaplan calculate are needed to rear a young human to independence, a mother needs help.

So how did our prehuman and early human ancestresses living in the Pleistocene Epoch (from 1.6 million until roughly 10,000 years ago) manage to get those calories? And under what conditions would natural selection allow a female ape to produce babies so large and slow to develop that they are beyond her means to rear on her own?

The old answer was that fathers helped out by hunting. And so they do. But hunting is a risky occupation, and fathers may die or defect or take up with other females. And when they do, what then? New evidence from surviving traditional cultures suggests that mothers in the Pleistocene may have had a significant degree of help — from men who thought they just might have been the fathers, from grandmothers and great-aunts, from older children.

These helpers other than the mother, called allomothers by

sociobiologists, do not just protect and provision youngsters. In groups such as the Efe and Aka Pygmies of central Africa, allomothers actually hold children and carry them about. In these tight-knit communities of communal foragers — within which men, women, and children still hunt with nets, much as humans are thought to have done tens of thousands of years ago — siblings, aunts, uncles, fathers, and grandmothers hold newborns on the first day of life. When University of New Mexico anthropologist Paula Ivey asked an Efe woman, "Who cares for babies?" the immediate answer was, "We all do!" By three weeks of age, the babies are in contact with allomothers 40 percent of the time. By eighteen weeks, infants actually spend more time with allomothers than with their gestational mothers. On average, Efe babies have fourteen different caretakers, most of whom are close kin. According to Washington State University anthropologist Barry Hewlett, Aka babies are within arm's reach of their fathers for more than half of every day.

Accustomed to celebrating the antiquity and naturalness of mother-centered models of child care, as well as the nuclear family in which the mother nurtures while the father provides, we Westerners tend to regard the practices of the Efe and the Aka as exotic. But to sociobiologists, whose stock in trade is comparisons across species, all this helping has a familiar ring. It's called cooperative breeding. During the past quarter century, as anthropologists and sociobiologists started to compare notes, one of the spectacular surprises has been how much allomaternal care goes on, not just within various human societies but among animals generally. Evidently, diverse organisms have converged on cooperative breeding for the best of evolutionary reasons.

A broad look at the most recent evidence has convinced me that cooperative breeding was the strategy that permitted our own ancestors to produce costly, slow-maturing infants at shorter intervals, to take advantage of new kinds of resources in habitats other than the mixed savanna-woodland of tropical Africa, and to spread more widely and swiftly than any primate had before. We already know that animal mothers who delegate some of the costs of infant care to others are thereby freed to produce more or larger young or to breed more frequently. Consider the case of silver-backed jackals. Patricia Moehlman, of the World Conservation Union, has shown that for every extra helper bringing back food, jackal parents rear one extra pup per litter. Cooperative breeding also helps various species expand into habitats in which they would normally not be able to rear any young at all. Florida scrub-jays, for example, breed in an exposed landscape where unrelenting predation from hawks and snakes usually precludes the fledging of young; survival in this habitat is possible only because older siblings help guard and feed the young. Such cooperative arrangements permit animals as different as naked mole rats (the social insects of the mammal world) and wolves to move into new habitats and sometimes to spread over vast areas.

What does it take to become a cooperative breeder? Obviously, this lifestyle is an option only for creatures capable of living in groups. It is facilitated when young but fully mature individuals (such as young Florida scrub-jays) do not or cannot immediately leave their natal group to breed on their own and instead remain among kin in their natal location. As with delayed maturation, delayed dispersal of young means that teenagers, "spinster" aunts, real and honorary uncles will be on hand to help their kin rear young. Flexibility is another criterion for cooperative breeders. Helpers must be ready to shift to breeding mode should the opportunity arise. In marmosets and tamarins — the little South American monkeys that are, besides us, the only full-fledged cooperative breeders among primates — a female has to be ready to be a helper this year and a mother the next. She may have one mate or several. In canids such as wolves or wild dogs, usually only the dominant, or alpha, male and female in a pack reproduce, but younger group members hunt with the mother and return to the den to regurgitate predigested meat into the mouths of her pups. In a fascinating instance of physiological flexibility, a subordinate female may actually undergo hormonal transformations similar to those of a real pregnancy: her belly swells, and she begins to manufacture milk and may help nurse the pups of the alpha pair. Vestiges of cooperative breeding crop up as well in domestic dogs, the distant descendants of wolves. After undergoing a pseudopregnancy, my neighbors' Jack Russell terrier chased away the family's cat and adopted and suckled her kittens. To suckle the young of another species is hardly what Darwinians call an adaptive trait (because it does not contribute to the surrogate's own survival). But in the environment in which the dog family evolved, a female's tendency to respond when infants signaled their need — combined with her capacity for pseudopregnancy — would have increased the survival chances for large litters born to the dominant female.

According to the late W. D. Hamilton, evolutionary logic predicts that an animal with poor prospects of reproducing on his or her own should be predisposed to assist kin with better prospects so that at least some of their shared genes will be perpetuated. Among wolves, for example, both male and female helpers in the pack are likely to be genetically related to the alpha litter and to have good reasons for not trying to reproduce on their own: in a number of cooperatively breeding species (wild dogs, wolves, hyenas, dingoes, dwarf mongooses, marmosets), the helpers do try, but the dominant female is likely to bite their babies to death. The threat of coercion makes postponing ovulation the better part of valor, the least-bad option for females who must wait to breed until their circumstances improve, either through the death of a higher-ranking female or by finding a mate with an unoccupied territory.

One primate strategy is to line up extra fathers. Among common marmosets and several species of tamarins, females mate with several males, all of which help rear her young. As primatologist Charles T. Snowdon points out, in three of the four genera of Callitrichidae (*Callithrix, Saguinus,* and *Leontopithecus*), the more adult males the group has available to help, the more young survive. Among many of these species, females ovulate just after giving birth, perhaps encouraging males to stick around until after babies are born. (In cotton-top tamarins, males also undergo hormonal changes that prepare them to care for infants at the time of birth.) Among cooperative breeders of certain other species, such as wolves and jackals, pups born in the same litter can be sired by different fathers.

Human mothers, by contrast, don't ovulate again right after birth, nor do they produce offspring with more than one genetic father at a time. Ever inventive, though, humans solve the problem of enlisting help from several adult males by other means. In some cultures, mothers rely on a peculiar belief that anthropologists call partible paternity — the notion that a fetus is built up by contributions of semen from all the men with whom women have had sex in the ten months or so prior to giving birth. Among the Canela, a matrilineal tribe in Brazil studied for many years by William Crocker of the Smithsonian Institution, publicly sanctioned intercourse between women and men other than their husbands — sometimes many men — takes place during village-wide ceremonies. What might lead to marital disaster elsewhere works among the Canela because the men believe in partible paternity. Across a broad swath of South America — from Paraguay up into Brazil, westward to Peru, and northward to Venezuela — mothers rely on this convenient folk wisdom to line up multiple honorary fathers to help them provision both themselves and their children. Over hundreds of generations, this belief has helped children thrive in a part of the world where food sources are unpredictable and where husbands are as likely as not to return from the hunt empty-handed.

The Bari people of Venezuela are among those who believe in shared paternity, and according to anthropologist Stephen Beckerman, Bari children with more than one father do especially well. In Beckerman's study of 822 children, 80 percent of those who had both a "primary" father (the man married to their mother) and a "secondary" father survived to age fifteen, compared with 64 percent survival for those with a primary father alone. Not surprisingly, as soon as a Bari woman suspects she is pregnant, she accepts sexual advances from the more successful fishermen or hunters in her group. Belief that fatherhood can be shared draws more men into the web of possible paternity, which effectively translates into more food and more protection.

But for human mothers, extra mates aren't the only source of effective help. Older children, too, play a significant role in family survival. University of Nebraska anthropologists Patricia Draper and Raymond Hames have just shown that among !Kung hunters and gatherers living in the Kalahari Desert, there is a significant correlation between how many children a parent successfully raises and how many older siblings were on hand to help during that person's own childhood.

Older matrilineal kin may be the most valuable helpers of all. University of Utah anthropologists Kristen Hawkes and James O'Connell and their UCLA colleague Nicholas Blurton Jones, who have demonstrated the important food-gathering role of older women among Hazda hunter-gatherers in Tanzania, delight in explaining that since human life spans may extend for a few decades after menopause, older women become available to care for — and to provide vital food for — children born to younger kin.

Hawkes, O'Connell, and Blurton Jones further believe that dating from the earliest days of *Homo erectus,* the survival of weaned children during food shortages may have depended on tubers dug up by older kin.

At various times in human history, people have also relied on a range of customs, as well as on coercion, to line up allomaternal assistance — for example, by using slaves or hiring poor women as wet nurses. But all the helpers in the world are of no use if they're not motivated to protect, carry, or provision babies. For both humans and nonhumans, this motivation arises in three main ways: through the manipulation of information about kinship; through appealing signals coming from the babies themselves; and, at the heart of it all, from the endocrinological and neural processes that induce individuals to respond to infants' signals. Indeed, all primates and many other mammals eventually respond to infants in a nurturing way if exposed long enough to their signals. Trouble is, "long enough" can mean very different things in males and females, with their very different response thresholds.

For decades, animal behaviorists have been aware of the phenomenon known as priming. A mouse or rat encountering a strange pup is likely to respond by either ignoring the pup or eating it. But presented with pup after pup, rodents of either sex eventually become sensitized to the baby and start caring for it. Even a male may gather pups into a nest and lick or huddle over them. Although nurturing is not a routine part of a male's repertoire, when sufficiently primed he behaves as a mother would. Hormonal change is an obvious candidate for explaining this transformation. Consider the case of the cooperatively breeding Florida scrub-jays studied by Stephan Schoech, of the University of Memphis. Prolactin, a protein hormone that initiates the secretion of milk in female mammals, is also present in male mammals and in birds of both sexes. Schoech showed that levels of prolactin go up in a male and female jay as they build their nest and incubate eggs and that these levels reach a peak when they feed their young. Moreover, prolactin levels rise in the jays' nonbreeding helpers and are also at their highest when they assist in feeding nestlings.

As it happens, male, as well as immature and nonbreeding female, primates can respond to infants' signals, although quite different levels of exposure and stimulation are required to get them going. Twenty years ago, when elevated prolactin levels were first reported in common marmoset males (by Alan Dixson, for *Callithrix jacchus*), many scientists refused to believe it. Later, when the finding was confirmed, scientists assumed this effect would be found only in fathers. But based on work by Scott Nunes, Jeffrey Fite, Jeffrey French, Charles Snowdon, Lucille Roberts, and many others — work that deals with a variety of species of marmosets and tamarins — we now know that all sorts of hormonal changes are associated with increased nurturing in males. For example, in the tufted-eared marmosets studied by French and colleagues, testosterone levels in males went down as they engaged in caretaking after the birth of an infant. Testosterone levels tended to be lowest in those with the most paternal experience.

The biggest surprise, however, has been that something similar goes on in males of our own species. Anne Storey and colleagues in Canada have reported that prolactin levels in men who were living with pregnant women went up toward the end of the pregnancy. But the most significant finding was a 30 percent drop in testosterone in men right after the birth. (Some endocrinologically literate wags have proposed that this drop in testosterone levels is due to sleep deprivation, but this would probably not explain the parallel testosterone drop in marmoset males housed with parturient females.) Hormonal changes during pregnancy and lactation are, of course, indisputably more pronounced in mothers than in the men consorting with them, and no one is suggesting that male consorts are equivalent to mothers. But both sexes are surprisingly susceptible to infant signals — explaining why fathers, adoptive parents, wet nurses, and day-care workers can become deeply involved with the infants they care for.

Genetic relatedness alone, in fact, is a surprisingly unreliable predictor of love. What matters are cues from infants and how these cues are processed emotionally. The capacity for becoming emotionally hooked — or primed — also explains how a fully engaged father who is in frequent contact with his infant can become more committed to the infant's well-being than a detached mother will.

But we can't forget the real protagonist of this story: the baby. From birth, newborns are powerfully motivated to stay close, to root — even to creep — in quest of nipples, which they instinctively suck on. These are the first innate behaviors that any of us engage in. But maintaining contact is harder for little humans to do than it is for other primates. One problem is that human mothers are not very hairy, so a human mother not only has to position the baby on her breast but also has to keep him there. She must be motivated to pick up her baby even before her milk comes in, bringing with it a host of hormonal transformations.

Within minutes of birth, human babies can cry and vocalize just as other primates do, but human newborns can also read facial expressions and make a few of their own. Even with blurry vision, they engage in eye-to-eye contact with the people around them. Newborn babies, when alert, can see about eighteen inches away. When people put their faces within range, babies may reward this attention by looking back or even imitating facial expressions. Orang and chimp babies, too, are strongly attached to and interested in their mothers' faces. But unlike humans, other ape mothers and infants do not get absorbed in gazing deeply into each other's eyes.

To the extent that psychiatrists and pediatricians have thought about this difference between us and the other apes, they tend to attribute it to human mental agility and our ability to use language. Interactions between mother and baby, including vocal play and babbling, have been interpreted as protoconversations: revving up the baby to learn to talk. Yet even babies who lack face-to-face stimulation — babies born blind, say — learn to talk. Furthermore, humans are not the only primates to engage in the continuous rhythmic streams of vocalization known as babbling. Interestingly, marmoset and tamarin babies also babble. It may be that the infants of cooperative breeders are specially equipped to communicate with caretakers. This is not to say that babbling is not an important part of learning to talk, only to question which came first — babbling so as to develop into a talker, or a predisposition to evolve into a talker because among cooperative breeders, babies that babble are better tended and more likely to survive.

If humans evolved as cooperative breeders, the degree of a human mother's commitment to her infant should be linked to how much social support she herself can expect. Mothers in cooperatively breeding primate species can afford to bear and rear such costly offspring as they do only if they have help on hand. Maternal abandonment and abuse are very rarely observed among primates in the wild. In fact, the only primate species in which mothers are anywhere near as likely to abandon infants at birth as mothers in our own species are the other cooperative breeders. A study of cotton-top tamarins at the New England Regional Primate Research Center showed a 12 percent chance of abandonment if mothers had older siblings on hand to help them rear twins, but a 57 percent chance when no help was available. Overburdened mothers abandoned infants within seventy-two hours of birth.

This new way of thinking about our species' history, with its implications for children, has made me concerned about the future. So far, most Western researchers studying infant development have presumed that living in a nuclear family with a fixed division of labor (mom nurturing, dad providing) is the normal human adaptation. Most contemporary research on children's psychosocial development is derived from John Bowlby's theories of attachment and has focused on such variables as how available and responsive the mother is, whether the father is present or absent, and whether the child is in the mother's care or in day care. Sure enough, studies done with this model in mind always show that children with less responsive mothers are at greater risk.

It is the baby, first and foremost, who senses how available and how committed its mother is. But I know of no studies that take into account the possibility that humans evolved as cooperative breeders and that a mother's responsiveness also happens to be a good indicator of her social supports. In terms of developmental outcomes, the most relevant factor might not be how securely or insecurely attached to the mother the baby is — the variable that developmental psychologists are trained to measure — but rather how secure the baby is in relation to *all* the people caring for him or her. Measuring attachment this way might help explain why even children whose relations with their mother suggest they are at extreme risk manage to do fine because of the interventions of a committed father, an older sibling, or a there-when-you-need-her grandmother.

The most comprehensive study ever done on how nonmaternal care affects kids is compatible with both the hypothesis that humans evolved as cooperative breeders and the conventional hypothesis that human babies are adapted to be reared exclusively by mothers. Undertaken by the National Institute of Child Health and

Human Development (NICHD) in 1991, the seven-year study included 1,364 children and their families (from diverse ethnic and economic backgrounds) and was conducted in ten different U.S. locations. This extraordinarily ambitious study was launched because statistics showed that 62 percent of U.S. mothers with children under age six were working outside the home and that the majority of them (willingly or unwillingly) were back at work within three to five months of giving birth. Because this was an entirely new social phenomenon, no one really knew what the NICHD's research would reveal.

The study's main finding was that both maternal and hired caretakers' sensitivity to infant needs was a better predictor of a child's subsequent development and behavior (such traits as social "compliance," respect for others, and self-control were measured) than was actual time spent apart from the mother. In other words, the critical variable was not the continuous presence of the mother herself but rather how secure infants felt when cared for by someone else. People who had been convinced that babies need full-time care from mothers to develop normally were stunned by these results, while advocates of day care felt vindicated. But do these and other, similar findings mean that day care is not something we need to worry about anymore?

Not at all. We should keep worrying. The NICHD study showed only that day care was better than mother care if the mother was neglectful or abusive. But excluding such worst-case scenarios, the study showed no detectable ill effects from day care *only* when infants had a secure relationship with parents to begin with (which I take to mean that babies felt wanted) and *only* when the day care was of high quality. And in this study's context, "high quality" meant that the facility had a high ratio of caretakers to babies, that it had the same caretakers all the time, and that the caretakers were sensitive to infants' needs — in other words, that the day care staff acted like committed kin.

Bluntly put, this kind of day care is almost impossible to find. Where it exists at all, it's expensive. Waiting lists are long, even for cheap or inadequate care. The average rate of staff turnover in day care centers is 30 percent per year, primarily because these workers are paid barely the minimum wage (usually less, in fact, than parking-lot attendants). Furthermore, day care tends to be age-graded, so even at centers where staff members stay put, kids move annually to new teachers. This kind of day care is unlikely to foster trusting relationships.

What conclusion can we draw from all this? Instead of arguing over "mother care" versus "othercare," we need to make day care better. And this is where I think today's evolution-minded researchers have something to say. Impressed by just how variable child-rearing conditions can be in human societies, several anthropologists and psychologists (including Michael Lamb, Patricia Draper, Henry Harpending, and James Chisholm) have suggested that babies are up to more than just maintaining the relationship with their mothers. These researchers propose that babies actually monitor mothers to gain information about the world they have been born into. Babies ask, in effect, *Is this world filled with people who are going to provide for me and help me survive? Can I count on them to care about me?* If the answer to those questions is yes, they begin to sense that developing a conscience and a capacity for compassion would be a great idea. If the answer is no, they may then be asking, *Can I not afford to count on others? Would I be better off just grabbing what I need, however I can?* In this case, empathy, or thinking about others' needs, would be more of a hindrance than a help.

For a developing baby and child, the most practical way to behave might vary drastically, depending on whether the mother has kin who help, whether the father is around, whether foster parents are well-meaning or exploitative. These factors, however unconsciously perceived by the child, affect important developmental decisions. Being extremely self-centered or selfish, being oblivious to others or lacking in conscience — traits that psychologists and child-development theorists may view as pathological — are probably quite adaptive traits for an individual who is short on support from other group members.

If I am right that humans evolved as cooperative breeders, Pleistocene babies whose mothers lacked social support and were less than fully committed to infant care would have been unlikely to survive. But once people started to settle down — ten thousand or twenty thousand or perhaps thirty thousand years ago — the picture changed. Ironically, survival chances for neglected children increased. As people lingered longer in one place, eliminated predators, built walled houses, stored food — not to mention inventing things such as rubber nipples and pasteurized milk —

infant survival became decoupled from continuous contact with a caregiver.

Since the end of the Pleistocene, whether in preindustrial or industrialized environments, some children have been surviving levels of social neglect that previously would have meant certain death. Some children get very little attention, even in the most benign of contemporary homes. In the industrialized world, children routinely survive caretaking practices that an Efe or a !Kung mother would find appallingly negligent. In traditional societies, no decent mother leaves her baby alone at any time, and traditional mothers are shocked to learn that Western mothers leave infants unattended in a crib all night.

Without passing judgment, one may point out that only in the recent history of humankind could infants deprived of supportive human contact survive to reproduce themselves. Certainly there are a lot of humanitarian reasons to worry about this situation: one wants each baby, each child, to be lovingly cared for. From my evolutionary perspective, though, even more is at stake.

Even if we manage to survive what most people are worrying about — global warming, emergent diseases, rogue viruses, meteorites crashing into earth — will we still be human thousands of years down the line? By that I mean human in the way we currently define ourselves. The reason our species has managed to survive and proliferate to the extent that 6 billion people currently occupy the planet has to do with how readily we can learn to cooperate when we want to. And our capacity for empathy is one of the things that made us good at doing that.

At a rudimentary level, of course, all sorts of creatures are good at reading intentions and movements and anticipating what other animals are going to do. Predators from gopher snakes to lions have to be able to anticipate where their quarry will dart. Chimps and gorillas can figure out what another individual is likely to know or not know. But compared with that of humans, this capacity to entertain the psychological perspective of other individuals is crude.

The capacity for empathy is uniquely well developed in our species, so much so that many people (including me) believe that along with language and symbolic thought, it is what makes us human. We are capable of compassion, of understanding other people's "fears and motives, their longings and griefs and vanities," as the novelist Edmund White puts it. We spend time and energy worrying about people we have never even met, about babies left in dumpsters, about the existence of more than 12 million AIDS orphans in Africa.

Psychologists know that there is a heritable component to emotional capacity and that this affects the development of compassion among individuals. By fourteen months of age, identical twins (who share all genes) are more alike in how they react to an experimenter who pretends to painfully pinch her finger on a clipboard than are fraternal twins (who share only half their genes). But empathy also has a learned component, which has more to do with analytical skills. During the first years of life, within the context of early relationships with mothers and other committed caretakers, each individual learns to look at the world from someone else's perspective.

And this is why I get so worried. Just because humans have evolved to be smart enough to chronicle our species' histories, to speculate about its origins, and to figure out that we have about thirty thousand genes in our genome is no reason to assume that evolution has come to a standstill. As gene frequencies change, natural selection acts on the outcome, the expression of those genes. No one doubts, for instance, that fish benefit from being able to see. Yet species reared in total darkness — as are the small, cave-dwelling characin of Mexico — fail to develop their visual capacity. Through evolutionary time, traits that are unexpressed are eventually lost. If populations of these fish are isolated in caves long enough, youngsters descended from those original populations will no longer be able to develop eyesight at all, even if reared in sunlight.

If human compassion develops only under particular rearing conditions, and if an increasing proportion of the species survives to breeding age without developing compassion, it won't make any difference how useful this trait was among our ancestors. It will become like sight in cave-dwelling fish.

No doubt our descendants thousands of years from now (should our species survive) will still be bipedal, symbol-generating apes. Most likely they will be adept at using sophisticated technologies. But will they still be human in the way we, shaped by a long heritage of cooperative breeding, currently define ourselves?

GARRET KEIZER

Sound and Fury

FROM *Harper's Magazine*

> In those days the world teemed, the people multiplied, the world bellowed like a wild bull, and the great god was aroused by the clamour. Enlil heard the clamour and he said to the gods in council, "The uproar of mankind is intolerable and sleep is no longer possible by reason of the babel." So the gods agreed to exterminate mankind.
>
> — the Epic of Gilgamesh

IN THIS WAY begins the earliest written version of the Great Flood myth, which reappears in altered form more than a thousand years later as the biblical story of Noah and the Ark. The Sumerian origins of the Gilgamesh epic may help to explain why the gods are so incensed. Sumeria is generally considered the world's first civilization; that is, the first place where human beings create a distinctly urban society. It seems that by at least the third millennium B.C.E. the world has begun to grow noisy, at least for those living in the mud-brick cities of the Fertile Crescent, the first people on earth to hear the shake, rattle, and roll of turning wheels. The city that never sleeps is born here, memorialized in a story about angry gods who cannot sleep either. The twenty-first-century tenant who some nights would like to strangle his noisy neighbor lives no more than a story up from some literate Mesopotamian who apparently imagined drowning his.

Human noise is political from its inception, not only because it emerges with the *polis* — that artificial forest where the tree that falls always makes a sound — but also because it lends itself so well to political conflict. Noise is both an objective and a subjective phenomenon; it comprises both common and uncommon ground. On the one hand, a decibel is a decibel is a decibel. The fact that the human ear can endure about two continuous hours of a power drill but only thirty minutes of a typical video arcade before sustaining permanent hearing loss and the related fact that eighty-year-old Sudanese villagers hear better than thirty-year-old Americans are just that: facts. On the other hand, the reasons why an airport will affect its neighbors in different ways, leaving some depressed or hypertensive and others relatively unfazed, are as variable and invisible as sound itself. Even the gods who confer with Enlil are not of the same party on the noise issue. At least one of them objects to eliminating human commotion at its source, and consequently a chosen few are able to step blinking from the ark into a temporarily quieter world.

By the time we get to the monotheistic universe of Genesis, the flooding of the earth is presented as the result of God's moral indignation. Wickedness, not noisiness, is what starts the rain. Yet I can think of few symbols more suited to wickedness than noise, usually defined as "unwanted sound" — like defining an assault as "unwanted attention." Loud noise hates nature and nurture alike. Certain species of birds fail to learn their mating songs, and therefore to reproduce, in noisy environments; as early as 1975, researcher Arline Bronzaft found that children on the train-track side of a New York public school were lagging a year behind their classmates on the other side of the building in learning to read. Even relatively low levels of noise can interfere with conversation (at 55 to 60 decibels); the price of making ourselves heard is a loss of nuance, inflection, vocal stamina — in every sense a "loss of voice." Noise has been linked to heart disease, high blood pressure, low birth weight, gastrointestinal disorders, headaches, fatigue, insomnia — in short, to nearly every known byproduct of stress. (Antistress medications are actually tested by exposing experimental subjects to loud sounds.) Noise deafens us, aurally and — there is strong evidence to suggest — morally as well. People subjected to high levels of noise are less likely to assist strangers in difficulty, less likely to recommend raises for workers, more likely to administer electric shocks to other human subjects.

Noise speaks danger; it both threatens and invites aggression. It triggers the physiological chemistry of the "fight-or-flight" response. Before we were even human, noise signaled the approach of the carnivore, of lightning and lava. More recently it became the alarm of invasion, first of the barbarian outside the gates and increasingly of the barbarian within. The audio-terrorist turns into decibels the dynamics of every relationship based on unrequited power: my noise can penetrate your quiet, but your quiet can never penetrate my noise. "My noise is my right" means "Your ear is my hole."

To this little rant of mine the Roman philosopher Seneca offers a censorious tut-tut. "I cannot for the life of me see that quiet is as necessary to a person who has shut himself away to do some studying as it is usually thought to be. Here am I with a babel of noise going on all about me." He is living over a bathhouse; the noises rising from downstairs include all manner of "grunting," "hissing," and "pummeling," as well as "the hair remover, continually giving vent to his shrill and penetrating cry in order to advertise his presence." Seneca boasts that he takes "no more notice [of] all this roar than . . . of waves or falling water." To be distracted by noise, he claims, is to succumb to one's own inner disquiet. I believe him, somewhat. But let us credit Seneca's stoicism for enduring the 50 decibels of a very loud hiss and estimate the cry of the hair remover at, say, 65 decibels, 75 if he has a very loud voice. Today Seneca would be living over a gym where the amplified music might be cranked to 100 decibels in order to produce the adrenaline rush that keeps the iron pumping. (And for every increase of 10 decibels, the volume of sound doubles.) If Seneca were one of the quarter million New Yorkers living within a hundred yards of an elevated train track, he might be able to test his fortitude with a screech of 115 decibels, as measured at the front step of his apartment building. At these levels philosophical detachment is almost a joke. What do you call a stoic who lives near the el? Deaf.

City ordinances aimed at noise date roughly from Seneca's time; Caesar is said to have banned nighttime chariot riding from the streets of Rome. Fiorello La Guardia tried to outlaw organ grinders; New Yorkers, to their lasting credit, told him to back off. Noise not only confers power; it is silenced by power. As anti-noise activists are quick to point out, the traditional noise ordinance has usually been aimed at and enforced against the individual. The kid with the boom box is one thing; the Federal Aviation Administration, which virtually regulates itself, is quite another. Indeed, the most effective noise ordinances I've found were gag agreements imposed as "compromises" on the critics of noise-making companies: in exchange for ninety "mufflered days" at the auto racetrack, your citizens' group will hereby withdraw its litigation; in exchange for an offer to purchase your soon-to-be-worthless house, you agree not to oppose the permit application of our soon-to-be-opened quarry. Ronald Reagan's shutting down of the Office of Noise Abatement and Control in 1982 and the failure of any president since to reopen it may go down as the most effective "noise ordinance" in American history. At the time there were eleven hundred local and state programs monitoring noise; now there are about twenty. Has anybody ever said, "Turn that damn thing off!" with greater success? One should always be wary, then, of equating quiet and silence. In the politics of quiet and noise, silence sides with the winner.

"I shall shortly be moving elsewhere," Seneca writes at the end of his essay on noise. "Why should I need to suffer the torture any longer than I want to . . ." Then, as now, stoic resignation was good, but an aggressive realtor was even better. From the first wheel that rattled through the streets of Sumer, the story of noise has always been tied to the story of human mobility — not least of all in America, arguably the world's first nomadic civilization. If there are any fundamental principles to the relationship between noise and mobility, I can discern at least two. The first and simplest goes like this: people move to escape noise, and by moving they always find it.

Forty years ago the poet Galway Kinnell went to northeastern Vermont looking for "a house with a view at the end of a long dirt road" that he could buy for $800. Outside the somewhat haggard village of Sheffield he found what he was looking for, or the closest approximation: a fallen-in farmhouse with broken windows and missing doors, which during Kinnell's earliest visits he shared with a pair of porcupines and a weasel. "Living out of reach of human activity," close to wild animals but within hearing of "the accent and pleasure of words, the love of getting things right" that Kinnell says "was more true of old-time Vermonters than any other people I know," the poet also found the materials for his work. And he found quiet, measurably more quiet on a summer evening than the sound of a human whisper, enough to live what he refers to as "an objectification . . . of my inner life." Of course, the danger in objectifying your inner life is that someone might drive a bulldozer over your heart.

The bulldozer arrived two years ago to clear-cut a nearby tract of land for the site of a South African–owned granite quarry. For many in town the appearance of the South Africans in Sheffield held a promise of jobs, tax revenues, and royalties, perhaps a less obscure place on the map. Kinnell had a different view. Fearful of what the newcomers might do to the landscape and suspicious of what they might have done to other landscapes and villages beforehand, he and a handful of like-minded neighbors opposed the quarry company's application for an environmental permit. Kinnell has already begun to hear the noise of blasting and heavy equipment in the distance and fears that he will soon see giant grout piles rising where he now sees nothing but trees and the chimney smoke of a few isolated neighbors.

About a year ago I began to follow the permitting process, which included a contested noise demonstration and a hilarious soggy trek through cedar swamp with lawyers wearing inappropriate shoes. The process held a special fascination for me, in part because Kinnell had seemed to be living the same dream that brought me to live in a town just one ridge over from his and was now living the nightmare that always dogs such a dream. I, too, count moose and bear among my nearest neighbors. I can see virtually every star one is able to see in the northern hemisphere. In the stillest hours of the early morning, Adam in his Garden has little over me. "It's so quiet here," say my guests from New Jersey. "So peaceful."

And so vulnerable. Quiet, after all, is the most assailable form of wealth. The same thief can forever be stealing it. It can grow back in a moment's respite, like the liver of Prometheus, only to be devoured by the screaming eagles once again. To tell a truth I seldom admit, sometimes I feel most at peace when I am seated on the porch of my in-laws' house in a blue-collar town in New Jersey, watching the kids play on the sidewalk and listening to the manhole cover tap amiably under the frequent traffic. Everything is settled there: I mean all the land and all the possibilities that haunt you when you're tempted to believe that your world extends as far as you can see and hear.

All of this is to locate the psychological place from which I began my exploration of quiet and noise. Like most quests, it began with "a passing sight," a disturbing image that one cannot easily suppress. Mine was the sight of Galway Kinnell's face when an environmental board member asked him, somewhere on the dirt road that led to the quarry, "Now, where is your house, Mr. Kinnell?" He turned and pointed toward a hill, several miles across an open field, with an obliging smile and the eyes of someone who has realized for the first time that he can die.

Again and again, the people I found in the forefront of some antinoise campaign were individuals who had moved from a noisier place to a quieter, only to have that quieter place grow loud. When Jane Moore came to Jerome, Arizona, some twenty-eight years ago, it was practically a ghost town. Formerly the site of a booming copper mine and home to fifteen thousand people, it was then a cluster of mostly abandoned buildings on either side of a desert highway. Moore was among the artists and squatters who began taking up residence next door to the few locals who had managed somehow to hang on. It was a place of steep canyons and weird acoustics: a clack of pool balls or a wind chime's jangle sounding in unexpected corners like the traces of restless spirits. For Moore, who had grown up next to a freight switching yard in Chicago, five miles away from O'Hare Airport, this was the place of quiet ambience she had always been searching for. She may find herself searching again. There are presently about three hundred fifty adults in Jerome, including three police officers. On some weekends as many as five hundred motorcycles pass through town, many sporting "modified" exhaust pipes that together with the terrain amplify the thunder of their descent through the canyon. A favorite stop is a rock-and-roll bar in town called the Spirit Room. Vice Mayor Moore and her associates in town government are attempting to pass a noise ordinance; the Modified Motorcycle Association of Arizona has promised that any such law will not go unchallenged. Bikers made the same point by driving into town one weekend and filling it with the modified sound of their presence.

There's an implicit cultural symbolism in conflicts such as this, none more pronounced than what I found in the ongoing feud between the New Hampshire International Speedway in the working-class town of Louden and the residents of scenic Canterbury, home to the "most intact and authentic of all Shaker villages" in America. The countryside and much of the architecture around Canterbury are about as arcadian and colonial as one could imagine. Some eighteen hundred of the original three-thousand-acre Shaker holdings, roughly seven hundred owned by the Shaker museum itself and the rest owned privately, are all under conservation easement. The tranquillity of the place would seem to be "secured," established, unassailable. But on major race weekends — as opposed to ordinary race weekends, which are virtually every weekend from April to October — the noise of 450-horsepower stock cars in neighboring Louden can reach the decibel level of jet planes. Standing in the front yard of a Canterbury residence, one sound expert measured noise levels as high as 85 decibels, roughly the same level as a lawn mower heard from six feet — this coming from a source half a mile away. For Hillary Nelson, who moved from New York to Canterbury with her husband, such noise is an attack on her quality of life. To many of the people in Louden, who have received tax revenues, a new ball field, a new fire engine, and $50,000 a year in college scholarships from the racetrack, noise is what you pay for quality of life.

In both Louden and Jerome, the source of the offending noise grew from a smaller, preexisting source of tolerable noise. As long as anyone can remember, there has been a biker bar in Jerome and a racetrack in Louden. What is more, by all accounts the patrons of both of these earlier establishments, though fewer, were wilder than the people who frequent them now. As the Noise Pollution Clearinghouse's Les Blomberg explained to me, one of the most frequent arguments made against those bothered by noise is that the offending noise source "was always there." The Sheffield quarry is being touted as a "reopening" of a pick-and-shovel operation that once extracted granite on the site. The Amcast plant in Cedarburg, Wisconsin, a large foundry that also has been at the center of a noise dispute with some of its neighbors, expanded from a small munitions plant during World War II to a major supplier of castings for the automobile industry. In each of these cases the disingenuous argument of "prior occupancy" is accompanied by avowals and sometimes even by hard evidence that the noisy party is trying, somewhat, to be "a good neighbor." Amcast, for instance, spent an estimated $400,000 in an attempt to make its foundry quieter, and seems to have succeeded. Even his critics credit racetrack-owner Bob Bahre with running an orderly operation that caters to a "family" clientele. Likewise, the owner of the Spirit Room bar is known to his patrons and to those who'd just as soon see his patrons spirited away for urging bikers not to rev their engines in town. Nevertheless, the thought of a ninety-thousand-seat NASCAR track being "a good neighbor" to a rural village, or five hundred motorcyclists becoming good neighbors to a town of four hundred, is something like the thought of King Kong being a good lover to Fay Wray. It may be sincere, it may even be noble, but if you're the one gripped in the big hairy paw it can only feel obscene.

The subject of noise and scale is of particular interest given the value we now place on cultural and biological "diversity." The soundscape provides both an example of diversity and an instructive analogy to other domains. A smaller sound can coexist with a number of other smaller sounds, but even a number of smaller sounds cannot coexist with one big noise. Never mind the forest — if a baby falls during a rock concert, does it make a sound? Different forms of the same question can be applied to species, languages, and ways of life. Those who dismiss the noise issue as "merely aesthetic" are, of course, ignoring the well-documented medical and psychological effects of noise. They are also forgetting that, in the context of relationships, aesthetics can become ethics.

The other thing that interests me about noise disputes is the way in which class conflict informs them, or seems to inform them. Of course it was no surprise to learn that a poorer life is frequently a noisier one, that those with low incomes are more likely to suffer from noise than the affluent, more likely to work next to the motor, more likely to live next to the airport, more likely to rent the apartment lower down and with thinner walls. But controversies between communities and a noisy industry are likely to pit neighbor against worker, at least in the eyes of the worker, who may tend to see the neighbor as someone with a good job who doesn't mind threatening someone else's job. This was certainly my impression when I was hanging out with the half-dozen Sheffield quarry workers while environmental experts, lawyers, and those with "party status" made their inspections of the site. "Maybe we'll get lucky and the whole bunch of them will fall into a hole," said a young equipment operator. Certain supporters of the quarry were quick to frame the issue as a conflict between the interests of the poet on the hill and those of the peon in the valley, between people who had come from elsewhere with money in their pockets and those who'd lived in the area all their lives with none. This is an old and bitter distinction in my neck of the woods, and one not without relevance to the politics of noise and quiet. When I first moved to "the north country," I often wondered why some of the men I talked to spoke so loud, until I realized that they had been partially deafened by millwork, chain saws, and tractors. I imagine that to many of these men, the idea of someone with an indoor job or a university education being sensitive to noise amounts to something like a personal insult, like holding your nose at the smell of your baggage carrier's sweat. And to approach things from the other side, I also imagine that certain displays of noise are intended as personal insults. Power in the hinterlands grows not only out of the barrel of a gun but also out of the barrel of an exhaust pipe and anything else that makes a good loud bang. Class warfare can come down to a war of sensibility: Fuck with me and I'll park something big and ugly across from your breakfast nook. Piss me off and I'll teach it to sing.

But the relationship between noise and class could be more peculiar than I supposed, and this became clearer in the case of recreational as opposed to occupational noise. What I discovered was how often the appearance of class struggle was manipulated to stereotype a dispute and how often that appearance was deceiving. That fracas over motorcycles in Jerome: obviously a fight between working-class guys having a little fun on the weekend and a bunch of potters and weavers living off trust funds, right? Not exactly. If you're looking for Marlon Brando, he ain't here. These days a new Harley-Davidson can cost nearly $20,000, and the typical "biker" in Jerome is an anesthesiologist or investment broker from Flagstaff getting in touch with his primal side. As for the track in Louden, a third of those who attend its NASCAR races have incomes of over $40,000 a year, and the drivers themselves need a hundred grand just to get a car ready for the track.

"They try to shape the battle into good old regular folks who like racing and these rich eggheads up on the hill," Hillary Nelson complains, and Bob Bahre was indeed quoted in the newspaper as saying that the sixteen Canterbury residents who had appealed a court ruling allowing him to expand his stadium capacity were "all wealthy people" who "just don't care about anybody else." This may not be as hypocritical a statement as it first seems; Bahre grew up on a hardscrabble farm, raced stock cars when the sport was strictly blue-collar in New England, and pretty marginal at that, and is still known for joining his cleanup crew in picking up litter off the race grounds. At some level, the Canterbury sixteen, who included a nurse, a salesman, and a musician, probably *did* seem like "the wealthy" to him. The fact remains that the money, the power, and the noise are on his side.

Back in Sheffield, I found myself taking the same second look. The "elitists" allied with Kinnell included two trailer folk who eke a subsistence living from the land, the sort of people who never count when small-town populists wax eloquent about "the people." The self-appointed defender of South African enterprise and local common sense who wrote letters to the newspaper identifying himself with the "peasants" against the "elitists" from outside was a transplant from New Jersey. The "native Vermonter" who complained about those people "who say, 'We've been to New York, we've had all the benefits of New York, but we don't want you to have them,'" had children with degrees from Stanford and Tufts — which doesn't exactly discredit her point of view but does suggest that the New York types have been a bit lax of late in maintaining their chokehold on cultural advantages. I became suspicious of any easy alignment of quiet and noise with privilege and deprivation. And I found that my own sentiments on the issue could shift as suddenly as sound on a windy day.

One Friday I followed the winding road out of Canterbury and abruptly found myself at the crowded intersection of the main highway through Louden at the start of a Winston Cup weekend. Entering a line of traffic that recalled evacuation scenes in disaster movies, I thought how stock-car racing could stand for everything I find distasteful in American civilization: the needless noise, the ubiquitous advertising, the waste of resources, the risking of human life for "special effect," the primacy of all things Caucasian and masculine, the out-of-shape motor culture's cherished belief that the best form of contact sport is one in which an athlete's buttocks make prolonged contact with a foam seat, the taking over of what was once the domain of inventive amateurs by the all-hallowed "pro" and his all-but-professional fans. I stopped to talk with one young enthusiast in the parking lot of an ice cream stand near the track. "What if they could make race cars quieter?" I asked him. "Would you go for that?" With the beatific smirk of a street evangelist declaring to every passerby that "God is love," he told me, "Noise is good."

But as I approached the track itself, I was not insensible of a certain magnificence, of the stadium rising like an immense, elliptical cathedral out of a sprawling metropolis of RVs and white billowing pavilions such as one might have seen at a jousting tournament. Carloads and truckloads of the faithful streamed in from the highway to pay homage to a Yankee farm boy's dream of building a world-class racetrack and a Tennessee farm boy's dream of being able to harness enough horsepower to outrun every revenuer on the road. Hundreds of banners proclaimed his victory, checkered flags and beer brands blazoned on every one, like lilies and crosses on Easter. Mister, you want to talk about myths and civilizations and the growth of ancient cities — this is our myth and our city, and we build it in two days, more than a hundred thousand of us in a swirling blur like pilgrims circling the Kaaba, exulting — as another race fan puts it — "in the rev and the roar."

I stopped counting after the third or fourth anti-noise activist told me that "noise pollution will be the secondhand smoke issue of the new century." I ought to have been happy to hear it. To say that I resent noise even more than I resent cigarette smoke is to say that I resent it very much. And yet I couldn't hear the comparison between noise pollution and secondhand smoke without wincing. Maybe I detect in the campaign against noise, as in the campaign against smoking, a flavor of that ruthless "progressivism" that first manifested itself when families of Neanderthals began to disappear oh so mysteriously from among their Cro-Magnon neighbors. "Must be the evolution," said the others, shifting their feet and whistling. As our society moves from a manufacturing base to an "information base," and as more and more blue-collar workers put on the livery of the service sector, is it any surprise that we should find our old machinery too noisy and the vices of those who tend it too intolerable, that we should demand our servants keep their voices down (like that Roman master so infuriated by unexpected noises that he had his slave thrown into a pond of lampreys for accidentally dropping a tray of crystal glasses) and take their nasty habits outside, while indoors we fuss to attain the perfect funereal quiet of an online chat room?

Might at least some of the noise assailing us amount to a protest against the threat of cultural or economic extinction? "I make noise, therefore I am." The Hispanic gardeners who recently went on a hunger strike to protest an L.A. ban on leaf-blowers said that the law was aimed at their race. In effect, they were saying that a noise *identified* them; silencing it was an attack on them. In a similar vein: "To everybody who told me I'd go nowhere in life: I can't hear you." That's a Sony advertisement for a car stereo system that cranks out sound at 164 decibels, loud enough to kill fish, but it easily could serve as the slogan for a generation that is not so much lost as unclaimed. Maybe the best way to fight the Boomers is with something that booms.

I wonder, too, if some of our antipathy toward noise isn't the negative form of our totalitarian consumerism, the belief that we ought to be able, as though by divine right, to achieve complete satisfaction of our every distaste no less than of our every desire. And will the ear ultimately lead the eye to more refined levels of fastidiousness? My interest in noise ordinances took me to an affluent New Jersey suburb where "hawkers, peddlers, and vendors," as well as "yelling, shouting, whistling, or singing on the public streets," are all prohibited by statute and where hanging up wash on an outdoor clothesline is prohibited by custom. It's funny to imagine that any people on earth, much less well-to-do people in the wealthiest nation on earth, would deny themselves or their neighbors the inestimable luxury of making love on sun-dried, wind-kissed sheets, or the wistful hope of some enchanted morning coming upon a peddler. If you met Humphrey Bogart in heaven, would you ask him to put out his cigarette? If you get to heaven at all, will you ask them to turn down the choir?

"The Lord is in his holy temple; let all the earth keep silence before him."

But this too: "Make a joyful noise unto the Lord, all the earth."

I hate noise as much as anyone I know, and I can flatter myself with the names of others in history who hated it, too: Darwin, Proust, Goethe, Poe, Haydn, Chekhov. Recent studies tend to confirm Schopenhauer's hunch: "I have for a long time been of the opinion that the quantity of noise anyone can comfortably endure is in inverse proportion to his mental powers." But I cannot forget that in addition to hating noise Schopenhauer hated the fact that he'd been born (and Poe felt that the most inspiring women were, shall we say, the extremely quiet kind). The connections worry me.

I would never want to forget what Thoreau said about the train: "when I hear the iron horse make the hills echo with his snort like thunder, shaking the earth with his feet . . . it seems as if the earth had got a race now worthy to inhabit it." Or what James Agee said about the importance of playing Beethoven *loud.* I agree wholeheartedly with the motto of the Noise Pollution Clearinghouse, "Good Neighbors Keep Their Noise to Themselves," and I try my best to practice it, but something else in me wants to cry, "Aw, go ahead" when Bob Marley sings,

> I want to disturb my neighbor
> Cause I'm feeling so right
> I want to turn up my disco
> Blow them to full watts tonight

True, whenever I take overnight accommodations, I always ask first if there are any wedding parties or traveling sports teams likely to "blow them to full watts" in the rooms nearby; yet few scenes in literature delight me as much as the one in Robertson Davies's *Fifth Business,* in which an anonymous Spaniard, having complained the night before about a row he mistakes for a boisterous honeymoon ("very well for you, señor," he calls through the door, "but please to remember there are zose below you who are not so young"), leaves flowers and this message at the door the next day: "Forgive my ill manners of last night. Love conquers all and youth must be served. May you know a hundred years of happy nights. Your Neighbour in the Chamber below."

To all of this, I have no doubt, many anti-noise activists would say, "We love these sounds no less than you. The problem is that every single one of them, Agee's quaint phonograph and Thoreau's equally quaint train, the joyful noise and the rub-a-dub style, is being overpowered by the boom car and the air horn, by a cacophony that is literally making us deaf" (and that Canadian noise expert Winston Sydenborgh estimates is doubling worldwide every ten years). Blake said, "No bird soars too high, if he soars with his own wings," but we are dealing now with the dark raptors of limitless amplification. Point granted. It was that secondhand smoke business that got to me, I guess. It was talking with Paul Miluski, the soon-to-be-without-his-lease owner of the Spirit Room biker bar in Jerome, whom I liked instantly, as I would have to like any man who plays croquet on a rooftop. Robert Frost said that he needed the poor for his work; I need a few Paul Miluskis for mine. And if they do not make at least some noise, how shall I know where to find them?

My noisy self-contradictions are very American, of course. Being an American is about living in contradiction, if it is about anything, because the glory and the tragedy of America come from our insatiable desire always to have the best of both worlds. That includes the worlds of noise and quiet, of utter freedom and inner peace. We want our own backyard version of the cloistered walk and our own Promethean stereo system as well. We want to practice Zen but mainly in the art of motorcycle maintenance. The purists, those who register one impulse or another as an enthusiasm and an allegiance rather than both impulses together as a complex form of yearning, are meeting now on the shrinking landscape and, perhaps more significantly, in the soundscape. Like radio stations on a crowded dial, their frequencies clash at certain bends in the road.

I actually take this meeting as a hopeful sign in that both of these impulses are essentially apolitical; that is, they both tend to express themselves against the quotidian sounds of the *polis*. Whether you choose to follow Huckleberry Finn or the Buddha, you always start by lighting out for the territories. Perhaps the exuberant noisemaker and the quiet seeker will discover that they are natural allies in spite of themselves, because each will of necessity have to appeal to the very sense of public domain and public life that once seemed anathema to their desires. To preserve either liberty or tranquillity against the passion of the other's counterclaim, one must in the end circle back to a rational discussion at the city gates.

The other reason I can take hope from the noise issue is its ability to penetrate and subvert political positions just as sound can penetrate — and, given the right Jericho, break down — a wall. Where would you locate the right and left wings of the noise pollution issue, for instance? Everywhere and nowhere. You can see noise as a threat to the most basic principles of private property, hearth and home. In other words, you can see noise as a threat to all those things that ought to be most dear to the conservative heart. Or you can see noise pollution as a threat to "the commons," an allusion frequently made by Les Blomberg and others in the anti-noise movement to the doomed English practice of preserving some common ground for community grazing. The soundscape is self-evidently property that no one owns, or rather that all of us own together. If it is possible to construct a "unified field theory" for our conflicting political currents, might it be found there? And if we could establish an ethic for sharing the soundscape, might that in turn pull us — by the ear, so to speak — to an ethic for sharing other forms of wealth? If any of these hopes is well founded, it may rest on nothing more sophisticated than the old wisdom of old neighborhoods, which says that the only sure way to hold a loud party without complaints from the neighbors, and with some hope of sleeping late and quietly the next day, is to invite all the neighbors to the party.

My hopes are probably not well founded, however. As we divide our world more ruthlessly into rich and poor, and the countryside into what Wendell Berry calls "defeated landscapes and victorious (but threatened) landscapes," it is probable that we will do the same thing in regard to sound. Many of us will live in pandemonium, and a few of us will have the means to live in paradise. The permeability of the soundscape may yet teach us to recognize the flaw in that arrangement, but we are likely to interpret the warning alarm as no more than a call to move elsewhere. Most of us will continue to put our faith in our mobility, in being able to run from the noise to someplace quieter. Like our pre-human ancestors, we still respond to noise with a fight-or-flight response, which at our stage of development means weighing the relative costs of the lawyer and the Realtor.

Of course, the irony of flight is the sound of flight itself. By far the largest number of noise complaints in this country have to do with modes of transportation, highway noise first of all, airport noise after that. Earlier I said that the first of two principles governing the relationship between mobility and noise is that people move to escape noise, and by moving they find it. The second principle is that people move to escape noise, and by moving they make it.

Along with fight and flight is there not a third ingrained response to the overwhelming power of noise, which is to fawn, to assume the position of joining what we cannot beat? So those who cherish quiet we dismiss as failed stoics, which may mean nothing more than that we are resigned to being cynics. In any case, I have begun to notice a curious thing about noise, which is how the pursuits disturbed or destroyed by it — pursuits such as writing a poem, watching a bird, or even looking after a child — can be made to sound so insignificant precisely because they make so little sound. Hillary Nelson told me of a day when her three-year-old son fell and hurt himself in her front yard, but she could not hear him crying over the drone of the race cars in Louden. I suspect that many will be as deaf to her complaint as she was to his cry. Kids fall all the time, right? From Jerome, Jane Moore wrote me a letter about a friend who moved out of town because she was dying of cancer and the motorcycle noise was making the process more painful. And I imagined the same cynical voice responding, "Let me get this straight. You need quiet to die?"

Actually, the time may be coming when you will not even need quiet to be dead. A German media artist who finds the notion of a quiet grave "idiotic" has recently created an exhibit of vocal tombstones, one of which "moans lustfully" when stroked. I find myself thinking about the moaning tombstone whenever someone tells me, with a faith so innocent it can bring tears to your eyes, how Technology (invoked with a capital T) is going to be our solution to the noise problem. Of course it can be, with marvelous results. In Europe, where noise reduction has a much higher place on the political agenda than it does here, roads are being built that reduce traffic noise by as much as 70 percent. Even the Harleys exported to Europe are designed to run quieter and, as an unintended result of such tinkering, turn out to have even more horsepower than their hoggish American cousins. Mechanical noise, after all, is an inefficient loss of energy (though many American consumers still equate louder volume with higher performance). So it is sometimes possible, with a little know-how, to have the best of both worlds. But for every noise we quiet, we produce another. Most of all, we continue to produce a false sense of virtual quiet by distancing ourselves from the actual noise we make. This is the ultimate form of "civilized" mobility: the removal of my actions from their effects. I don't have to hear the printing presses that publish my words, the strip-mining equipment that feeds my computer. I'm a writer, you see. I practice a quiet occupation. Technology has carried that old suspicious adage about not shitting where you eat to a place perilously above suspicion, where highly intelligent people are capable of believing that they don't even shit where they shit. When I tried to suggest as much to some of the movers and shakers in the noise movement — for instance, to the consultant for an airport-noise group who told me that he logs seventy-five thousand flight miles a year in his work — the response I got was often a bit chilly. Perhaps this was due to the understandable fear that someone already in danger of being branded a crank might also be branded a Luddite. But perhaps it rather had to do with a deal already struck with the successful manufacture of the first mud brick some nine or ten millennia ago. As the consultant told me, "The absolutist approach [i.e., the one I had just proposed to him] says we must change the way we live . . . but no one's going to stop the growth of airlines. Why tilt at windmills you can't defeat? I wouldn't want to stop the evolution of our science, even though there are going to be losers."

So in the end the most raucous disturbance of the peace (or the most cynical response to the complaints of those disturbed) may be nothing else but the brazen form of a more discreet and universal communication, the signed version of an anonymous chain letter that the rest of us mail out every day and that comes back with interest to everybody's mailbox sooner or later. (And sooner than later will be able to announce its arrival by moaning lustfully.) When this essay is done, for example, I will send it to New York City by overnight mail because it absolutely positively has to be there, and because the other pursuits of my have-it-all life will undoubtedly push the project too close to its deadline. Overnight mail is only possible with overnight flight, which has made no small contribution to the 2,156 percent increase in air-cargo traffic since 1960. Sometime in the night the plane carrying my little meditation on noise will fly over someone's roof, waking her or her aged father or her colicky three-month-old from a sound sleep. Wrapping her robe furiously around herself, perhaps going so far as to light a cigarette with her trembling fingers, she will curse that plane, and, to some small degree that I cannot gauge and can never hear, she will be cursing me. And those cranky old gods of Sumer and Uruk, long since deaf if not dead, will do no more than I will to help.

VERLYN KLINKENBORG

The Pursuit of Innocence in the Golden State

FROM *The New York Times*

I REMEMBER HER vividly, a woman standing in the liquor aisle of a grocery store in a Sacramento suburb. She had a cigarette in one hand and with the other she placed a fifth of store-brand vodka in the grocery cart. I was fourteen, and it was a summer morning, 1966. My family had moved to California from Iowa only a week before, fresh as a load of turnips just lifted from the soil. I knew the woman was divorced by the way she held her cigarette, by the way she did not look around to see if anyone was watching when she took the vodka off the shelf. In Iowa, liquor was segregated in a state-owned store with all the sensory appeal of an auto-body shop. No one divorced, and smoking was, if nothing worse, a waste of matches.

Iowans moved to California in droves about then, and not to drink, divorce, and smoke. They went for all the reasons people still go to California, for jobs and a less punishing climate and a new beginning. But one of the thrills of California, like seeing a line of palm trees or eating an artichoke or an avocado, was coming across a woman like that, a Tippi Hedren who converted what were vices in one state into the ordinary domestic manners of another and who implied, by her bearing, that if there was a problem, it was your problem. I look at Lisa Yuskavage's dewy, bosomy paintings, and I remember that California seemed exactly that voluptuous to me, its sins undone by the frankness with which they were revealed. It was easy to believe that California was somehow defined by the casual license that woman embodied. But the woman, shopping for vodka, as it were, in her negligee, and the boy looking on, entranced and scandalized — that is the true social and ethical dynamic of the golden state. It has always replicated, in its own way, the dynamic of America itself, a nation where some came to make a fresh start with a new rigor and others came to free themselves from rigor. If it could make electricity with the turbine of desire and disapproval, California would never run out of power.

Nowadays, no one dreams of lighting a cigarette in a California supermarket, and even the store-brand vodkas try to look as upscale as the Finnish and Swedish brands. Divorce is the common coin of modern sexuality, as the children of many of those Iowa-raised Californians grew up to discover for themselves. I remember how surprised I was to see farms when my family moved to California, when hops fields still grew on the southern outskirts of Sacramento. But California has been doing away with that unpleasant surprise for the past three decades. Now there are houses and houses and houses.

Electricity is a metaphorical form of energy as well as a literal one, a symbol, ever since the first kilowatt was generated, of human power over nature. There are, of course, complicated, interlacing causes for California's current energy crisis, rooted in bad planning and bad deregulatory designs. But one cause is the lack of human power over human nature, a growth in consumption fueled by an endless faith in repeated fresh starts, industrial and domestic. If you add up all of California's gospels — the canons of social responsibility and individual liberty, of market freedom and unfettered growth — you come up with a cumulative sense of innocence, which is one of the things that really gall the other Western states, which have come to understand California better than it understands itself.

Californians talk about nothing but change, because that is the landscape they live in. The rate of change is so rapid, and has been for so many years now, that the idea of conservation seems almost futile. Or rather, it seems commodified and impersonal, dependent on more efficient appliances, less consumptive houses, and cleaner cars instead of altered behavior and wiser choices. California believes in innocence with the same fervor it believes in newness, in the psychology of starting over again. Electricity is the perfect metaphor for that easy faith. It is always new, always fresh, intangible and killing at the same time.

ROBERT KUNZIG

Ripe for Controversy

FROM *Discover*

DRIVE SOUTHEAST from Dijon in France, across the broad plain of the Saône River, until the land starts to roll and the Jura Massif looms gray on the horizon, and you're in the Franche-Comté — Pasteur country. The great Louis was raised in Arbois, a small town nestled at the base of the Jura cliffs. You can still visit his ivy-covered childhood home, where his father tanned hides and where in later life Louis installed a bathtub (one of the first in town) as well as a laboratory in which to pass his summer holidays. On shelves in the lab stand the flasks of chicken broth — made in 1883 and untouched since — with which Pasteur disproved the theory of spontaneous generation. In the opposite corner is the sauna-like room where he tested the effects of heat on microbes. It was Pasteur, of course, who discovered that you could kill microbes by heating them — or "pasteurizing" them, as people would later call it.

A few hundred yards from that shrine, in the center of Arbois, you can also watch cheese maker Thierry Bobillier make Comté, the Swiss-style cheese that is the pride of the region. Clad in white from cap to rubber boots, in a workshop that is all tile and gleaming stainless steel, Bobillier dips his hand into a six-hundred-gallon vat of warm, butter-yellow liquid — the whey — and pulls out a fistful of white curds, like plump grains of rice. He squeezes them into a ball, stretches them like modeling clay, rubs them between his fingers; now they're ready to be pressed into five ninety-pound wheels of Comté. Ask Bobillier what he thinks of pasteurization and you get a hearty laugh. "We're not for it," he says. By French law, he can't make Comté from anything but raw, unpasteurized milk.

Cheese is milk that has been curdled and fermented by microbes. Fermenting microbes are naturally present in raw milk, but pasteurization wipes them out along with the pathogens. In industrial countries today, most cheese is made from pasteurized milk, to which the cheese maker has then added back a few selected microbial cultures to do the fermenting. But artisans like Bobillier, of whom there are many in France and a growing number in the United States, continue to make cheese from raw milk, arguing that its rich natural microflora adds flavor to the cheese. In the United States, pasteurization is required for all fresh or soft-ripening cheeses, but it is still legal to use raw milk for hard cheeses such as cheddar that are aged for at least sixty days.

The Food and Drug Administration, however, is thinking of changing that. Worried by experimental evidence that certain strains of pathogenic bacteria — such as the notorious *Escherichia coli* 0157:H7 — may survive even sixty days of aging, it is considering an outright ban on raw-milk cheese. The American government has already pushed for this position in negotiations for international food standards — the so-called Codex Alimentarius — that are supposed to smooth international trade. It has met, naturally, with vigorous resistance from France, land of four hundred cheeses, the finest of which are made of raw milk.

When you eat such a cheese, you are eating an evolving ecosystem. There are billions of bugs in every bite. In a Comté, as in most cheeses, the evolution begins with bacteria that convert lactose — milk sugar — into lactic acid. This sours and curdles the milk: the protein suspended in it, called casein, clumps together to form solid curds that float in the liquid whey. In many cheeses the curdling is hastened by adding rennet, an enzyme preparation taken from the fourth stomach of calves (or made synthetically). Calves curdle their mother's milk so that it can pass through their intestines slowly enough for the nutrients in it to be absorbed. Some five thousand years ago, humans started making cheese for practical reasons, too: they found it takes longer to rot than raw milk does. The good microbes help keep the bad ones in check (although before Pasteur no one knew about microbes).

Meanwhile, those lactic-acid generators foster other microbes that thrive in a sour environment and ripen the cheese, giving it taste and aroma. "You've got bacteria, yeasts, and surface-growing molds," says biochemist Pascal Molimard of Degussa BioActives, which sells cultured microbes to cheese makers. "Those microorganisms break down the proteins and lipids in the curd and use them as nutrients. And so the result is a large number of very small molecules which, being small, are volatile and pass into the atmosphere. It's the mixture of all those molecules that makes up the aroma of the cheese."

The mold *Penicillium camemberti,* for instance, is what gives a Camembert its mottled white rind; it sends enzymes into the interior of the cheese that break down the caseins into smaller molecules, which in turn propagate back to the surface and nourish the mold. In the process the interior of the cheese gradually changes from chalky paste to creeping ooze, and the Camembert, especially a raw-milk one, acquires a fungal aroma reminiscent of old tennis shoes. A closely related mold called *Penicillium roqueforti,* using similar enzymes, achieves a completely different result: the blue color and tangy "blue" aroma of Roquefort.

Companies like Degussa have cultured a few dozen of these microbes; an ambitious maker of pasteurized cheese might add five or six to the blank slate that is pasteurized milk. "But in a raw-milk cheese, it's not five or six strains you're adding, but a hundred," says microbiologist Eric Beuvier of the National Institute for Agronomic Research lab in Poligny, just down the road from Arbois. "Which makes for a great diversity of flavors." Different microbe strains may break down milk proteins in subtly different ways, producing different precursors of aroma molecules; more strains make for a more complex and distinctive aroma and flavor. Beuvier's team has shown that raw-milk cheeses contain larger and more diverse populations of bacteria than do pasteurized cheeses; tasting panels consistently distinguish raw-milk Comté because of its more powerful and pungent aroma — an aroma that so far only nature can provide.

Of course, *E. coli* 0157:H7 and *Listeria monocytogenes* are part of nature, too. Yet there is next to no evidence that these bugs have caused illness through aged raw-milk cheeses, as they do through contaminated meat. "If you look at the epidemiologic data, there are very few outbreaks that involve cheese at all," says food microbiologist Catherine Donnelly of the University of Vermont. "Particularly the aged hard cheeses appear to be microbiologically safe.

Even though cheese consumption is going up, [outbreaks from those cheeses] are just not showing up in the epidemiologic literature."

Ironically, the cheeses that have caused illnesses have often been made from pasteurized milk and then contaminated during processing. "Pasteurization may actually create a more dangerous situation, in that you knock out the competitive flora," Donnelly says. The good bugs that help keep the bad bugs in check in a raw-milk cheese are destroyed by pasteurization.

Donnelly did her study for the Cheese of Choice Coalition, an organization of "farmstead" cheese makers and cheese importers who have joined forces to resist mandatory pasteurization. France, they point out, has managed to avoid mass death-by-cheese by implementing that other great idea of Pasteur's: strict hygiene, both on the farm and at the cheese maker's. "There is a lot of rigor to the work that goes into this product," cheese maker Bobillier told this reporter, who at the time was wearing plastic booties, a gown, and a hairnet. Indeed, French raw milk has gotten so microbe-free of late that government scientists like Beuvier worry that it may be compromising the flavor of cheese. That concern may seem quaint to American authorities, but after all, cheese is supposed to be more than nonlethal. It's supposed to taste good as well.

ANNE MATTHEWS

Wall Street Losses, Wall Street Gains

FROM *Orion*

BY TWO in the morning, New York is as quiet as it gets. You can walk for blocks up Broadway and hardly see a moving car; you can stand at the corner of 42nd and Fifth and, sometimes, smell the sea. By three in the morning, the planet's most profoundly developed real estate has nearly shimmered back into its earliest self, forty rocky islands set in river, bay, and sound: Rikers, Swinburne, Black Bank, Plum. North Brother, and Castle, and Cuban Ledge. The Isle of Meadow. Ellis, and Coney. Long Island, where Brooklyn elbows Queens. Staten. Manhattan.

Of New York's five great boroughs, only the Bronx is part of the North American continent, and it contains both New York City's worst slums and its best stand of virgin timber. Manhattan is the most densely populated New York borough, Staten Island the most rural, Brooklyn the most populous. Queens is the largest in area and the most ethnically varied, even in a city where a third of the residents are foreign-born.

By four in the morning, each member of this urban archipelago sails alone. To go directly from Wall Street to Staten Island now, you almost need a kayak, or a canoe; the New York airports are silent, the commuter ferries rock at pierside, and the river tunnels have nearly emptied too, their fluorescent lights fizzing peach and ice white and spring green fifty feet below the Hudson, churned by twice-a-day tides from sweet to salt. Tropical butterflyfish, gold and silver, black and white, are dozing beneath the city's piers and pilings, their colors dimmed as they rest. Warm Atlantic currents sweep coral-reef species into the New York estuary every year. Sea turtles, some the size of hubcaps, others half the size of full-grown steers, can be in the harbor too, usually dragged in by tankers. These do not care for city life, so the local Coast Guard has become proficient at turtle rescue and release. The Hudson is a blackwater river, with all the underwater visibility of chocolate milk, and a loud one, especially in winter, when snowflakes striking the water's surface set up a high-frequency roar. Human senses miss the snow-thunder entirely, but for eel and sturgeon, the great river is as noisy as Times Square.

Throughout the urban night, New York's 722 miles of subway line stay open, though service slows. Here an A train snakes east through the Jamaica marshes, there a late run noses toward the Columbia campus through its caves of white Inwood marble and gray Fordham gneiss. But the big roadways into the city are nearly deserted at this hour. Toll-takers nod in their booths beside the New York exit ramps of the Garden State Parkway, or the empty EZ-Pass lanes flung across the Tappan Zee. Even in Van Cortland Park, crossed by three Bronx expressways, the flying squirrels have emerged to dine and socialize, soaring from branch to branch under the municipal moon.

By five in the morning, some nights, the moon is down, sunk beyond the benzene inlets and cyanide pools of the Jersey Meadowlands west of town. The gently radioactive hills of Staten Island, veined with uranium-rich red and green serpentinite, are dwarfed by the dark bulk of the Fresh Kills landfill at the island's south end, at five hundred feet tall the highest point on the Eastern Seaboard. A few commercial fishing boats are still moving up the harbor, glimmering belowdecks with bluefish and cod for the city's five thousand restaurants; in urban bakeries from Canarsie to Turtle Bay, the yeast is popping and working in the dough. Though dance clubs and emergency rooms and newsrooms have stayed wakeful, most of New York's 8 million humans lie unconscious in their rented burrows, the city's dominant daylight species finally, grudgingly, asleep. Only the skyline blazes.

In the New York financial district, between midnight and dawn, security guards patrolling near the World Trade Center watch the night sky above Manhattan's tip and listen for birdsong. Billions of migrating birds rush over North America twice a year, seeking breeding grounds and winter homes, heading north with the spring and south in fall. Nearly a hundred species pass directly over Manhattan Island. Some of these long-distance commuters like to call to one another as they fly: white-throated sparrows heading from Honduras to breeding grounds in Quebec, magnolia warblers making the run from Panama to the Adirondacks. When the seasons are changing, you can stand on Wall Street in the small hours and hear the migrants calling, faint and high, as they stream above the sleeping city. Some travel singly, some in groups: a kettle of hawks, a siege of herons, a wedge of swans. Aerial traffic rises near each equinox, but migrating birds fly over Manhattan nearly every night in the year.

At six in the morning on this raw October Saturday, the financial district is deserted, and cloud wraps the twin corporate towers of the World Trade Center and the World Financial Center from base to crown. "I've brought fresh mealworms," Rebekah Creshkoff assures me, patting a side pocket in her khaki vest as we turn onto a silent Vesey Street. During the work week, Creshkoff tries to carry mealworms in her purse, just in case, but she probably will not use them today. Last night was foggy, and she expects the worst.

"Slink along the walls, so we don't startle anything," Creshkoff warns, scanning a concrete walkway beside the American Express headquarters. False alarm: wet cigar. False alarm: banana peel. First live sighting: a male cardinal perched in a potted yew, looking sleepy and cross. On the dank pavement beyond is a small still form. "Blackpoll," says Creshkoff softly, bending down — "no, black-and-white warbler." She spreads its wings with a fingertip, turns the creamy breast skyward to examine the ebony stippling, and peers at a still-lustrous dark eye. Then she seals the body in a plastic Ziploc bag, scribbles on it the date and place of discovery, and tucks the dead warbler deep in another vest pocket.

Rebekah Creshkoff is a latecome birder who grew up in New Jersey, went to Brandeis and Sarah Lawrence, and then (years into a business career) took a birding class on whim. She liked it enough to enter a Columbia University certification program in conservation biology. Creshkoff knows the financial district's glassy maze by heart because she works here, as a corporate communications officer at the Chase Manhattan Bank. But often she leaves her Upper West Side apartment at 5:45 A.M., biking eight miles down-island to check in with porters and doormen and security patrols, who tell her what they have seen in the night. Manny at the World Trade Center is especially vigilant.

"He's been picking up injured flyers for years," says Creshkoff, waving to a massive bundled figure. "He feels sorry for them. A big kid, and such tiny birds." Manny's tips, as always, are to the point. Stunned bird, near a brokerage entrance. Dead bird, on the Vesey Street sidewalk. Dazed and frantic bird, trapped under a glass overhang; seeing a ficus tree in the lobby, it apparently tried to roost but smashed into the building's window wall instead, confused by multiple reflections from wet marble and shadowy panes.

We trace a looping path around the bases of World Trade Towers One and Two, looking for crash victims. Rats the size of guinea pigs chitter to one another as they search the corporate lawns for injured songbirds to devour. "I found a scarlet tanager trapped in that revolving door once," says Creshkoff, pointing. "I put it in a vest pocket, called my office to say I had a dentist's appointment, and took it on the subway to Central Park." Liberated at the 59th Street entrance to Central Park, the tanager vanished into the treetops.

When Creshkoff does find stunned birds in the financial district, she coaxes them into a paper bag, carries them to shelter (an atrium garden, a corporate tree), then offers fresh mealworms and second chances. If she encounters a survivor, she can't keep it; under the federal Migratory Bird Act of 1916, you can be in possession of a live bird for twenty-three hours, but not twenty-five.

Mostly she finds the dead. The bright lights of office towers seem to short-circuit the natural navigational abilities of birds in flight. Sophisticated city breeds like pigeons and sparrows stay calm when they see a city skyline at night. Songbirds travel late, when air currents are calmest, and steer by the stars. But Manhattan buildings can be a quarter-mile high. Migrants see lights directly in their flight path, follow them trustingly, then circle the Chrysler Building or the World Trade Center, mesmerized, until exhaustion claims them.

In Toronto, thanks to a special plea by Prince Philip and the World Wildlife Fund, many hotel and office towers now turn off lights during migration season, even though it means relocating night workers to interior offices, rewiring lights, and sending employees out each morning to rescue fallen warblers. Chicago has begun a similar program, on a smaller scale. Hearing of the Toronto efforts, Creshkoff wanted to go to Manhattan building managers and ask their help directly. Better to collect hard evidence, the Canadians told her. So casualties retrieved on Creshkoff's rounds find temporary storage in her apartment freezer. When there's no more room for the Häagen-Dazs, she FedExes a load of frozen birds to researchers at Maryland's Patuxent Avian Research Laboratory. They send her a special cooler; she fills it with skyscraper kill and ships it back; they pay.

"It's hard to imagine New York's commercial landlords voluntarily dimming their lights at night," says Creshkoff gloomily. Though you never know — there's closet birders everywhere." (The Empire State management will sometimes dim their building in migration season, a welcome exception.)

We slink on, veering toward the brokerages nearer the harbor, checking the overpasses between buildings, the empty sidewalks, and the Winter Garden, whose handsome expanses of lighted glass are a prime deathtrap for birds. Creshkoff points again. Hopping at the edge of a steam grate is a young female yellowthroat (a neotropical migrant, my Peterson's field guide notes: summers in Canada, winters in the West Indies).

"The yellowthroats are tiny but resilient," says Creshkoff, watching it explore the gutter, head cocked. "Over half will live, once trapped among these buildings. Most songbirds won't."

Dawn is finally here, the chill half-light turning from slate gray to pearl all around us, the corporate towers vanishing halfway up into sea fog. Creshkoff's early-morning tours have brought all sorts of wild encounters. One spring, she found a live female red-bellied woodpecker clinging to a polished marble wall; and once a little brown bat huddled on a steel pillar, dazed with cold. She has found a dozen stranded woodcocks, as well as a Virginia rail which wandered the World Trade plaza for nearly a week, subsisting on french fries.

On her worst day ever, Creshkoff logged sixty-four birds, all dead. A good day, always, means capturing survivors and getting them to open space. World Trade victims are usually discovered in the margins and shadows, fatally baffled by stone, steel, and glass. On a Marriott walkway, we find a black-throated blue warbler with a broken neck; it probably flew to its death trying to reach the reflected trees in the hotel's window wall. "The black-throated blues are extremely vulnerable to city lights," Creshkoff says. "I've never found a live one."

As we cross the barren plazas, looking, knots of homeless men have begun to stand and stretch. The great towers have begun to warm up, too, and make small creaks and chirpings as they do, disconcerting to a birder. Soon this plaza will close for the winter, before the annual ice falls can begin. A hard winter in downtown Manhattan sends ice slabs dropping into the World Trade courtyard from 110 stories up. Halfway across the largest plaza, we spot a white-throated sparrow and a hermit thrush wandering the concrete reaches in short puzzled flights. "Hurry, guys," Creshkoff pleads. The peregrine falcons who nest on a window ledge near Wall Street will soon begin their morning hunt.

Six World Trade Center is the last set of glass walls between fallen migrants and the New Jersey hills, but in a dim corner we find a white-throated sparrow, several days dead.

From a concrete planter comes a sudden frantic chirring.

"Oh, my God, a winter wren! Jesus Christ! Be careful, boy!" The wren, small, dark, and round, darts down a pillared concourse and is lost to sight. We follow. No sign of the wren, but slumped in a dark corner is a dead female yellowthroat, her body still quite warm. The yellowthroat's wing feathers ripple for an instant in the Hudson River wind. Sighing, Creshkoff digs for a Ziploc.

"I could put her in an envelope and mail her with one first-class stamp. Feel how light."

Another Wall Street loss. I look up at the great towers with the clouds about their knees, and stroke the honey and amber of the dead breast. If I had not seen the warbler lying in my hand, I would swear my palm were empty.

When historians look at New York, they see an overgrown port town that exists for just one reason: making money. When sociologists look at New York, they perceive two cities: the workaday outer boroughs have far more in common with Rust Belt capitals like Baltimore or Cleveland than with theatrical, ravenous, self-centered

Manhattan. When human geographers look at New York, they see a viewshed (the places where Manhattan's skyline is first visible on the horizon, night and day). Or they map the city's news-shed, within which the *New York Daily News* is the morning's first read and not, say, the *Philadelphia Inquirer,* and sometimes its sports-shed as well — the zip codes where Yankees and Knicks are considered home teams.

Physical geographers studying the metropolitan area prefer to learn its watersheds, the areas of land that drain rainwater and snowmelt into the nearest marsh or lake or stream. Though New York is a strikingly energy-efficient city — mostly because it stacks and packs its residents, then makes them use mass transit — the New York suburbs invented sprawl, and sprawl makes floods, bad ones. Turbodevelopment has erased 90 percent of the New York region's original marshes and meadows, and the replacement fields of asphalt and concrete neatly repel water instead of letting it soak in.

When political scientists look at the New York conurbation, they see one of the great unnatural wonders of the policy world. The New York area is the most elaborate, least manageable civic aggregation in human history, a polycentric supramegalopolis. New York ignores the nation at its back whenever it can. It would like to be a city-state, but in four centuries has devised neither a governing authority nor (unlike New England or Southern California) a convincing regional narrative.

In the last fifty years, New York has outgrown at least three tries at definitive labeling. The postwar boom saw it evolve from an Industrial Age metropolis with one eye on Pittsburgh and St. Louis into a world capital of culture, finance, and communication whose real peers were London and Paris. By the mid-1950s and early 1960s — New York's Augustan age — urbanologists declared New York the star of Megalopolis, the six-hundred-mile skein of development from Boston to Washington that knits the Eastern Seaboard into one long, thin, supercilious supercity. Now urbanologists have begun to call New York our prime example of the galactic city. A galactic city (a term invented by geographer Peirce Lewis of Pennsylvania State University) is a tissue of development so vast that it creates its own order, in a burst of edge cities and technoburbs. In the galactic city, suburbs and exurbs no longer push outward from an urban core like rings on a tree. Instead, most expeditions and interactions are suburb to suburb; you create your own metropolis, measuring distance in travel time, not in miles from some distinctive central feature like Times Square or the Loop. Everyone's galactic-city map is different: the favored supermarket is ten minutes away, the preferred mall thirty minutes in another direction, the workplace forty minutes distant — urbanism à la carte.

But to an ecologist, New York is most interesting as an ecotone, a place where natural worlds collide — northern and southern climate zones overlapping, land meeting ocean, salt water mixing with fresh. Six natural habitats define the City of New York: estuary, salt marsh, woodland, beach, freshwater river, and prairie. Some of these ecosystems are relics now, like the improbable patch of virgin forest in upper Manhattan's Inwood Hill Park, and some are remakes, like the recreated eastern grassland at Brooklyn's old Floyd Bennett airfield. Centuries of human assault on New York's natural underpinnings have fragmented and degraded all the city's original habitats, sky and water, leaf and stone.

The earth scientists, especially, believe you cannot know a place without ground-truthing it. Ground truthing is the act of walking a piece of ground, or flying low over it, or rowing the waters around it, taking time to see as well as look, rather than trusting what tradition says is there, or what theory tells you should be. Manhattan is indeed the world's most densely developed real estate, as New York and its outliers are the very paradigm of sprawl. But even in the ultimate city — especially in the ultimate city — what you see depends on where you stand.

All five New York boroughs retain places where you can walk for hours and see no human near. Northern Manhattan has its lonely salt marshes; the Staten Island Greenbelt, three times the size of Central Park, shelters spring-fed kettle ponds and dwarf-pine forests. The Brooklyn Grassland's gentle prairie by the bay, punctuated by an abandoned control tower, supports 140 acres of wildflowers and big bluestem, a scene straight out of Kansas. Beside New York's largest airport, JFK International, lies the Jamaica Bay Wildlife Refuge, some parts all marsh, others sand and cactus. Snowy owls come to Jamaica Bay when the weather turns cold, eager to hunt rats and rabbits along the runways next door. The level winterscape of the great airport apparently reminds them of their native tundra in Siberia and Baffin Bay; whole clans of snowy owls now fly down from the Arctic each year to winter in Queens.

When a snowy owl looks at New York, it sees safety. And lunch. And a frontier. A great many birds and animals are discovering that city living can be less stressful than a career in the wild. Natural time scales may differ from ours, but nature's agenda never changes. It will take over, if it can. New York is the best possible place to start. Ecologists know that big cities are far more friendly to wildlife than small ones, because the potential habitat is both immense and varied. Parks and greenways and suburban gardens offer ideal hiding places and travel corridors; urban creeks and backyard lap pools and corporate fountains yield reliable fresh water. To a twenty-first-century raccoon or deer, New York (or Atlanta, or Frankfurt) looks like a fine big animal sanctuary, with the prime food sources in the middle of town.

Sometimes the incoming species are only taking over parts of the city that we avoid. Of the nation's twenty-five largest cities in 1950, eighteen had lost population to the 'burbs by 2000, the vertical city succeeded by the horizontal city. Urban researchers call it the paradox of the green ghetto, best seen in depopulating urban settings like the modern ruins of downtown Detroit, where pheasants fly over the aging freeways; in East Saint Louis, where weeds cover railroad tracks and the basics of city life — a corner store, a taxicab, a fire department — are rare sightings; in North Philadelphia, where small trees have begun to grow in the streets. New York is unusual because it still has a lively downtown, making it one of very few U.S. cities with both a vital heart and an active edge. But overall, the American city, in the last fifty years, has become more crowded and voracious at the margins, quieter and greener and wilder at the core.

After twenty years spent in and around America's largest city, I have begun to notice odd alterations in the texture of daily life here, little slubs in the weave. A cornfield appeared on Upper Broadway Avenue: a Dominican immigrant had noticed a nice piece of land going to waste, there in the median strip, and decided to farm it; the city let him. When a distinctly under-the-weather fox visited the inner suburbs, its cityward progress was breathlessly chronicled on

New York's all-news stations ("to see what a fox looks like, especially a rabid one, go straight to our Web site at www.1010.WINS!"). By 1999, coyotes and wild turkeys had begun to roam Central Park ("How did they get there?" demanded the *Wall Street Journal.* "Crosstown bus?"). By 2000, black bears had visited Chappaqua and the Palisades Parkway. White-tailed deer came back to Manhattan for the first time in generations, making late-night dashes down the Amtrak trestle at the tidal strait called Spuyten Duyvil, on the island's north end, where Henry Hudson once came ashore.

And between Newark and the Jersey Meadowlands one winter morning, I spotted from my train window a dozen egrets, flying low above the dank chemical mudflats, an arrow of white headed straight for the World Trade Center. What are they doing here? I wondered, horrified, amazed. How do they live? But a quarter century of water cleanup has brought ibis and yellow-crowned night herons and the shy and solitary bittern back to that former open sewer, New York Harbor. Hundreds of herons now breed on uninhabited islands off the Bronx and Queens. To see them, you must crawl ashore through great tangles of poison ivy, then hold up a truck mirror to observe their secret rookeries — but they're there, and flourishing. I had no idea.

It was a figure-ground problem, really. For years, I had looked at Greater New York and seen only what I expected to — a profoundly unnatural landscape; a competitive maze; a wonder of money and art that seemed a thrilling human triumph some days and on others a declensionist's delight. New York attracts jeremiads. Emerson called it a sucked orange, Fitzgerald pronounced its grimy suburban sprawl "the ugliest country in the world," Vonnegut thought it a skyscraper national park. Yet above, around, behind, below, I began to find another New York, suppressed or silent in daylight, exceedingly lively from twilight to dawn.

The next decades will be the first truly urban period in human history. At the turn of the twenty-first century, half of us were concentrated in the world's metropolitan areas, particularly twenty or so emerging supercities, chief among them New York, Los Angeles, London, Rio, Mexico City, Calcutta, and Nairobi. By 2050, three-fourths of our species will be city creatures. Already, one American in fifteen lives in New York, or in the New York suburbs.

Yet throughout the United States, as from Toronto to Tokyo,

nature/culture confrontation is becoming part of urban, suburban, and periurban routine. Some encounters charm us; some we dread; others we badly misunderstand. New York has long cultivated an edgy relationship with nature, that big green blur between the lobby and the cab. To be vague or dismissive about the natural world is the last acceptable prejudice in The City, which talks a lot about diversity, but about biodiversity hardly at all. For centuries now, the City of New York has resolutely rushed ahead, determined to find the best deal, to never waste time, to never show weakness. It rarely looks around, rarely looks back. Maybe it should. Wild does not always mean natural; urban is not the same as tame. Even in Manhattan, you are never more than three feet from a spider.

STEVE MIRSKY

Dumb, Dumb, Duh Dumb

FROM *Scientific American*

THE NEED for improvement in our nation's math and science education is a standard sentiment of our times. Indeed, a close scrutiny of recent news headlines, combined with a personal experience, indicates to me that our nation's math and science skills truly have plummeted to a value of *x*, where *x* is some number that is very, very low.

For example, consider the story of four young men who busted into a veterinarian's office in Noblesville, Indiana, in late August. The ne'er-do-wells were nailed after stealing what they thought was a painkiller known as OxyContin, which has gotten press lately because some idiots snort it to achieve a heroin-like high. Our callow dopes, however, apparently have an attention span of only three letters, for what they stole was in fact oxytocin, which helps females give birth, produce milk, and develop nurturing feelings toward their progeny. As the editor of a major American scientific magazine said after I told him about the confused criminals, "Maybe I'm wrong, but you've got to think that four young guys with enlarged, tender nipples and a tendency to cuddle are not going to fare that well in prison."

Just a few days before the aforementioned arrests came another example of the challenges faced by those who possess an IQ of *x*, where *x* is some number that is very, very low. This case concerned a Long Island woman who allegedly decided to end her marriage to her millionaire husband the old-fashioned way — by killing him. The flaws in her plan, however, were more fatal than the plan itself. An aide at a nursing home, the woman told her husband she needed him to help her practice drawing blood. She would, therefore and henceforth, regularly be sticking needles in his arm. But unbeknownst to him, she was shrewdly using dirty needles smuggled out of the nursing home, in the hopes of giving him AIDS.

How greatly she might have benefited from a sound science education. For one, AIDS is not exactly rampaging through nursing homes, so the odds of her bringing home a needle carrying HIV were slim. For another, it is extremely rare to get infected with HIV even after being stuck with a needle that has been in contact with HIV-positive blood: the transmission frequency is only about 0.3 percent. The woman, who merely succeeded in giving her husband more common and easily transmissible conditions, such as hepatitis, was caught after she ran out of patience and tried to hire a hit man to expedite matters. The hit man turned out to be a police informant, and the woman and her husband are now, one might say, legally separated.

Finally, also in late August, I found myself stuck in southbound traffic on the infamous elevated Bruckner Expressway in the beautiful Bronx. This traffic jam was special, as it consisted in large part of people who were ignorant, or at least apathetic, about mathematics. They were returning from Connecticut, which was selling tickets for the $295 million Powerball lottery, to New York, which does not. A few days later I expressed my frustration to Michael Orkin, professor of statistics at the California State University at Hayward and author of *What Are the Odds? Chance in Everyday Life.* He e-mailed back, "If you have to drive ten miles to buy a Powerball ticket, you're sixteen times more likely to get killed in a car crash on your way than you are to win." Share this statistic with any of the geniuses on the Bruckner, and they might say, "But we weren't on our way. We were on our way *back.*" Besides, with the road so clogged, any crashes would have occurred at a survivable x miles per hour, where x is some number that is very, very low.

JUDITH NEWMAN

"I Have Seen Cancers Disappear"

FROM *Discover*

HE WALKS with the hitch of John Wayne, his lanky six-foot-four frame lurching slowly but purposefully down the hospital corridor. He is tough, God knows, but he's no cowboy. His measured gait is the result of a massive tumor growing in his pelvic region; on his chest wall lies another mass the size of a grapefruit. The baseball cap worn at a jaunty angle covers a bald head, scarred from surgery to remove a growth in his brain. His skin is opalescent, like the meat of a coconut: cancer has blanched him. At fifteen, he was diagnosed with Stage IV metastatic melanoma. All conventional treatments had failed by the time he arrived here, at the NIH National Cancer Institute (NCI) in Bethesda, Maryland. He had, at most, three months to live.

That was three years ago. He is still here. Frustrated that he hasn't been able to join his friends in college and furious that occasional seizures have kept him from driving. But nevertheless here. At some point, if you're dying long enough, it begins to seem an awful lot like living.

Steven A. Rosenberg, his doctor, watches him warily. "This boy, he's been through a lot," Rosenberg says quietly. "At home they did a lymph-node resection, treated him with alpha interferon. The tumor came back. He came here, we gave him interleukin-2; he responded transiently. The tumor came back. We gave him newer treatments, very aggressive. The tumors kept growing. Now we've tried something brand-new — something we've tried on only twelve patients. I don't know. I don't know. We'll see what's happening today. But he's looking so well."

So well? A newcomer looking at the boy sees a perversity of nature, a child shadowed by death. Rosenberg sees a long-term survivor with a flicker of energy, who has miraculously put on twenty-five pounds in a month and may be responding to the most daring treatment the NCI has to offer. It may save him. Then again, it may kill him. What's it like to know that your efforts can tip that balance?

That, in short, is the dilemma that Steve Rosenberg, chief of surgery and head of tumor immunology research at the NCI for the last twenty-five years, grapples with every day. It's hard to believe that this unprepossessing, avuncular man sporting Gepetto eyeglasses and polyester geekwear is a man who still dreams about changing the world, but there it is: at the age of sixty, he continues to have faith that the cure for at least some cancers will be discovered in his lifetime and that he may be the one who makes the breakthrough.

On one side of Rosenberg's brain lives this fact: despite all the effort poured into cancer research, mortality rates have hardly budged. Higher five-year survival rates are misleading. Cancers are being detected earlier, which is probably why more people make the five-year mark. Looking at ten-, fifteen-, and twenty-year survival rates, for most cancers there is really no difference.

Then there is the other side of Rosenberg's brain. Here the dream is luminous and real. He knows that cancer can be beaten. "I have seen with my own eyes patients with widely metastatic, invasive, bulky cancers that would result in their deaths in a few weeks . . . I have seen these cancers disappear completely with immune manipulations." For every extraordinary success — remission — there are thousands of failures. Experimental cancer treatments at the NCI are reserved for those who have exhausted every other possibility; by definition, anyone who ends up under Rosenberg's care has been told he's got three months left. How does he soldier on in the face of such odds?

Steve Rosenberg was the last child born to Harriet and Abraham Rosenberg, Orthodox Jews who came to this country from Poland in the 1920s to escape the pogroms. As a boy he listened carefully to the family legends, at the heart of which lay great suffering and helplessness. Abraham had seen his father and a brother blown to pieces, shelled by the Germans in World War I. Then his mother fell ill with fever, and he wheeled her in a cart from hospital to hospital, trying to find one that would take a Jew. Turned away everywhere, she died by the side of the road, Abraham holding her in his arms.

When Rosenberg's parents reached this country, they settled in the Bronx, with no education to speak of but a belief in its transformative powers for their children. The elder Rosenberg awoke at 5:30 A.M. each day to open his luncheonette, grinding out dollars that would send his sons to medical school. Abraham spent Sundays studying the haftarah with his rabbi. "There was a phrase from the ancient rabbis that my father taught us," Steve Rosenberg says. "'He who studies his lessons a hundred times is not to be compared to he who studies his lessons a hundred and one times.' There was always something else to learn, or something you could learn more thoroughly."

Steve was only seven years old when a startling thought occurred to him: he had been born to fight evil. Not as a superhero, as a seven-year-old might imagine, but as a doctor. The idea solidified in his mind after his father began receiving the postcards. They began, "We regret to inform you," and Abraham would read about another death — an uncle, an aunt, a cousin killed by the Nazis. Almost everyone in his family was wiped out, including a beloved brother, who had escaped to France, only to be shot by the Nazis, leaving five children behind. His father would read the notes, and there would be silence. "I would hear about another death," Rosenberg remembers. "I would see my father and mother suffering, and I would just wonder . . . how could people act like that?"

The effect of these losses cannot be underestimated. Today, when Rosenberg talks about cancer, his metaphor is the Holocaust: "I've always been struck by the horror and tragedy of cancer — not only for the people who have it but also for their families, who have to stand by, impotent, and watch their loved one decimated. That was one of the horrors of the Holocaust too — the helplessness."

The world the Rosenbergs inhabited in the 1940s seemed out of control. Evil was random, unpredictable. For no good reason, houses were bombed, hospital doors shut in your face, families annihilated. Yet his mother and father carried on. They did not rage; they did not break down. This was a magnificent lesson for a budding scientist: learn to tolerate constant losses and accept them as integral to future achievement.

In his youth, Rosenberg imagined a rational, predictable world where he might control the uncontrollable — and what is more uncontrollable than cancer? — through a combination of brainpower and technique. Laboratories and surgical theaters: both little worlds unto themselves where people are helped and everything can be managed. As an adult, he would no longer be a passive witness. He would change things.

Rosenberg was rare in this way. He loved medical school and thrived on four hours of sleep a night. After interning in surgery at Peter Bent Brigham Hospital in Boston in 1963, he detoured residency, against the advice of his advisers, to earn a Ph.D. in biophysics at Harvard, struggling to find a way to mesh his desires to be in both the operating room and the lab. The number of doctors who manage to become surgeons and researchers is very small. It's like asking someone to function as a jockey and as a center for the NBA: there isn't a whole lot of overlap in skill sets. Everything about research requires waiting, observing, delaying gratification. Surgery demands action, split-second decisions. Rosenberg started out in surgery, but he always knew that to get where he wanted to go, surgery would not be enough. He has framed in his office a quotation from the eighteenth-century surgeon John Hunter: "Surgery is like an armed savage who attempts to get that by force which a civilized man would get by stratagem."

During these student years he developed a ferocity about his work that sometimes drove other people away. He was so consumed that when he finally met a woman he loved — Alice O'Connell, head nurse in the Brigham ER — he broke up with her repeatedly, telling her he couldn't abide the distraction. Alice hung in, he eventually puzzled out a way to allow himself the luxury of a private life, and they married.

In 1974, after eleven years of residency and fellowships, Rosenberg was offered the position of chief of surgery at the Dana Farber Cancer Institute in Boston, an almost unheard of offer for a man so young. He quickly accepted the job, then quickly turned it down when the NIH called with a better offer. That offer can be summed up by the setup of his office: exit through one door and walk into laboratories; exit through another door and walk into a ward full of patients. Here research could be translated into human medicine. It was what he craved.

*

As a resident, Rosenberg had determined that cancer was the biggest problem he could imagine tackling. Anything less would have been a failure of nerve, a refusal to fight the greatest evil.

This much he knew. Cancer is many different diseases. But the sine qua non of cancers is the uncontrolled growth of cells that do not know when to die. The mutant cells multiply and travel throughout the body, crowding out normal cells, rendering organs useless. Rosenberg knew that the problem researchers had to solve was this: How is it that cancer, like a vicious foreign virus or bacteria, can kill us when, unlike the virus or bacteria, it is part of us?

In 1968 he had the first inkling that the human immune system might hold the key. He was working at the veterans' hospital in West Roxbury, Massachusetts. A grizzled sixty-three-year-old man came in complaining of abdominal pain. Rosenberg found that he needed his gallbladder removed. But the man's medical history posed a mind-boggling question. Why was he still alive? Twelve years earlier, he had come in complaining of pain that turned out to be a huge malignant tumor in his stomach. The cancer had also hardened his lymph nodes and infiltrated his liver. The surgeon opened him up, saw the tumor was inoperable, and sent him home to die.

Yet here he was, complaining about his gallbladder.

When Rosenberg operated, he felt around everywhere in the man's abdominal cavity, looking for signs of tumor. There was nothing but soft, pink, healthy tissue.

Spontaneous remissions of metastatic cancer are not unknown, but this was the first time Rosenberg had actually seen one himself. This was also the time when the intersection of cancer and immunology was being discussed everywhere, when researchers were just figuring out that there were T cells, a class of white blood cells that attack foreign material in the body. When a foreign invader enters the body, marker compounds (called antigens) on the foreign substance act as red flags waved in front of the bull-like white blood cells. Once the T lymphocytes, a part of the white blood cell system, recognize the antigens as foreign, they start to attack the foreign cells. The horror of cancer is that the body fails to recognize cancer cells as foreign — or at least not foreign enough for the immune system to mount a full-scale assault. Exactly why the cells do not set off alarm bells is one of the enduring mysteries of science. What is clear is that if immune therapy is ever to work, the T cells must learn to recognize antigens on the tumor and signal the immune system to mount an attack. Theoretically, that's what happens in cases of spontaneous remission.

Rosenberg was obsessed by the case. He and his wife, Alice, took long walks together, discussing the whys and ifs of the veteran's recovery. He transfused another cancer patient with blood from the cancer survivor, but the second patient died. He tried endless variations, infusing different types of immune cells into sick patients. Eventually, he concentrated on renal-cell cancer and malignant melanoma because they were the first cancers that showed a response to immune manipulation.

There were many failures. Rosenberg is not alone among cancer researchers in this regard, but because he was bridging the roles of researcher and clinician, his experience was more personal. He was often there, standing at the side of the bed, when the experiment failed.

Death is something Rosenberg and Alice spend a lot of time thinking about. (For five years, she was director of nursing at the Whitman-Walker AIDS Clinic. Last year she moved to the NIH, where she is nurse case manager for AIDS patients.) "We often joke about what a fun couple we are," she says. "He does cancer; I do AIDS. Want to spend the evening with us?" But, she adds, "Talking about death and dying in my family is very natural. Like talking about the weather, or the Redskins. Not that it's belittled, it's just . . . that's what we do. We help people who are very sick, and sometimes we're not successful, and then they die. I think there's something people don't understand. Helping people to live is a privilege — there are a lot of people today who wouldn't be alive if Steve didn't do what he did. But helping people to die is a privilege too. To do it well, to help people die with dignity . . . It's the last real favor you can do for somebody."

Failure only made Rosenberg work harder. He spent many sleepless nights at the lab monitoring experiments. He tried to eat dinner with Alice and his three daughters. But when an experiment wasn't going well or a patient he'd had high hopes for wasn't responding, he might just as well have stayed at the hospital, a reality his family came to understand. "My father would be home on Sunday, wanting it to be a family day, and he'd just be pacing," says Rachel Pomerantz, twenty-seven, founder of mdexpert.com, a Web site that helps cancer patients find second opinions. "And we'd all be, like, *go, go back to work!*"

Despite his early disappointments, Rosenberg was the first to recognize a critical characteristic of T cells. "Steve recognized the importance of T-cell growth factor [interleukin-2] to immunology," says Mike Lotze, who worked in Rosenberg's lab and is now director of protein and gene therapies for GlaxoSmithKline. Interleukin-2 is a molecule that activates lymphocytes to attack tumor cells. Synthesizing it in their lab, Lotze adds, "gave us the ability to culture T cells, clone them, and grow them in vitro."

The year 1985 changed everything. As Ronald Reagan's surgeon, Rosenberg was the man who had to stand up at a White House press conference and say, of Reagan's colon polyps, "The president has cancer." So he became a known figure on the national scene. But he was about to become really known.

For three years, Rosenberg had been injecting patients with interleukin-2, or mixtures of interleukin-2 and LAK (lymphokine-activated killer cells), which are white cells transformed into tumor-killing cells by exposure to interleukin-2. During these hard years patient after patient died, sometimes in utter misery. Before he knew how to control the side effects, patients had to be treated in the ICU for chills, fever, joint pain, nausea, fluid retention, and liver failure or kidney problems. At one point he lost sixty-six patients in a row. "For a long time," says Rosenberg, "we were working in a dark room, not knowing whether we would ever find a door out of that room."

It began to look as if that door had been opened when he treated a series of patients with monstrous doses of interleukin-2 and LAK. Many responded; Linda Taylor, a twenty-nine-year-old naval officer with metastatic melanoma, saw her tumors melt away in six weeks. Fifteen years later, she is disease free.

In December 1985, he announced his results in the *New England Journal of Medicine:* twenty-five Stage IV cancer patients treated, regression of tumors in eleven — a 44 percent rate of response. Reporters swarmed; Rosenberg found his face on the cover of *News-*

week; and the NIH cancer hot line began receiving hundreds of calls a day.

By 1988, he and his team had published a study about the efficacy of tumor-infiltrating lymphocytes, killer T cells that don't just circulate in the bloodstream but make a beeline for the tumor. These cells aren't all that effective on their own, but when extracted from the tumor, fortified with interleukin-2, and reinfused into the body, they caused significant shrinkage in 40 percent of metastatic melanoma patients.

Shrinkage, not cure.

Whether Rosenberg hyped his results or whether the media hyped his results for him is a matter of debate among some researchers, but there is no question he became a one-man publicity machine for the NCI. Reading the press hosannas, one would think Rosenberg had single-handedly found the magic bullet for cancer. Rosenberg himself knew the results were modest. NCI studies showed that interleukin-2 alone (LAK cells eventually proved a red herring) can effect about a 9 percent cure rate in melanoma and renal-cell cancer patients — modest, but better than certain death.

Efforts to replicate those results have not succeeded thus far. A study published in 2000 by Franz Porzsolt at the University of Ulm in Germany and Chris Coppin at the University of British Columbia suggests that patients with Stage IV renal-cell cancer who are treated with interleukin-2 show an overall response rate of only 3.2 percent: better than the spontaneous remission rate, but not much. "This study's results may be accurate, but you have to put them in context," says Sloan-Kettering's Larry Norton. "First, the important thing is proof of the principle — that interleukin-2 works at all. Then, it is universally true that when a drug is tested at a single institution, it does better than in multi-institutions. The reason is patient eligibility. When the drug gets out, the people who get treated are not always the best candidates for the treatment." So it's not surprising, Norton insists, that subsequent results were not as good as those Rosenberg obtained.

Critics who resent the media attention showered on Rosenberg in the 1980s argue that his work of the last fifteen years has been more hype than substance. That he is not exactly a man to shun a microphone also irks some colleagues. "Oh, he's at the NCI, he's a very good speaker, he presents well — he's a very bright guy," says

Donald Morton, cancer vaccine pioneer and medical director and surgeon-in-chief of the John Wayne Cancer Institute. "He has always gotten a lot of visibility for anything he did — whether he was the first to do it or not."

But even naysayers agree with Rosenberg's supporters on a key point: he has been a tireless cheerleader. Because he has stood up for the principle of immunotherapy, billions of dollars are being spent to prove that the immune system can cure cancer. "When most people were trying to come up with the next generation of cytotoxins [chemotherapy drugs], Steve was talking about immunotherapy," says Michael Lotze. "He convinced people there were real, credible opportunities."

In recent years Rosenberg has spent great energy refining his techniques for manipulating interleukin-2 and tumor-infiltrating lymphocytes. Incubating the cells is an arduous process that must be individualized using each patient's cells, so it is unlikely a drug will ever arise from this method. In the first instance of true gene therapy for cancer, Rosenberg and French Anderson, who is now director of the gene therapy lab at the University of Southern California, introduced cancer-fighting genes into the tumor-infiltrating lymphocytes; then the cells were infused with a gene for tumor necrosis factor, a hormone that interferes with tumor blood supply. Out of twelve patients, only one has responded for any length of time. "We can insert genes into lymphocytes, but we can't control their gene expression," Rosenberg says. "If the cell turns off the genes we put in, we're out of luck."

Most recently he has been working, like many other researchers, on so-called cancer vaccines. The idea is this: antigens, those flag-waving fragments on the surface of cancer cells, are isolated from a patient's tumor. Injected into the body, the antigens trick the immune system into producing a flood of tumor-infiltrating lymphocytes, which then, at least in theory, go after the tumor cells containing the antigens. Out of thirty-one melanoma patients who received both interleukin-2 and the vaccine, which delivers a synthetic peptide that mimics the tumor antigen, thirteen saw their tumors shrink at least 50 percent after three series of injections. The combination is being tested at thirteen medical centers across the country.

Whether this approach will prove more successful than other

immunotherapeutic variations is a matter of debate among cancer researchers. Fearing immunotherapy may prove an endless and disheartening trail, many researchers are now traveling different roads that show promise.

Rosenberg is house-proud, and the NCI is his house. He gives a tour, but then stops to visit a few patients. He tries to sound upbeat without striking a false note; he sees the worst of the worst every day and knows it doesn't help anyone, himself included, to be a drama queen. "Not feeling too great, huh?" he asks a woman who can only answer with her eyes as she vomits into a bedpan. "You'll feel better soon."

While he is quick to mention the miraculous cures, he's even quicker to mention the failures. "The problem is, when I lie in bed thinking at night, it's not about the people who are better. I think about all the people who didn't make it." It's this sense of time passing, time wasted — and perhaps the fact that he himself has only a limited number of years left to work — that haunts him. This is also why the trend toward confidentiality and nondisclosure agreements at biotech companies infuriates him. "A major pharmaceutical company wants me to come to a meeting," he says. "They offer me $25,000. But they ask me to sign a statement saying I won't divulge anything I learn at this meeting for ten years. I won't sign it because that inhibits progress. Almost weekly I run into problems with this in my own research — where I need something, and companies won't send it unless I agree to keep the research a secret!"

It's not only other companies that infuriate him. "I once had a scientist come into my office, sit in that chair, and say, 'I've got some really exciting results to show you. But listen, it's really hot, and you've got to promise me you won't tell anybody about it.' Horrifying! I said, 'Of course I won't agree to that!' He wouldn't show it to me. We're supposed to be fighting the disease, not each other. There is no role for secrecy in science."

His position is the moral one; it's also one he can afford to take. As head of the 120-person surgery branch of the NCI with its $13 million budget, he is not in the trenches scrabbling for funding or academic advancement. "I've taken a lot of hits because of the visibility I've had and because of my position on this subject. Part of the reason, I guess, is jealousy. And the idea that scientists should be cloistered — I don't understand that at all. We have an obligation to help people understand what medical research is about."

What of the accusation, frequently hurled his way, that he has a surgeon's personality in a researcher's suit — that he can overlook human suffering because he is focused on the bigger picture? Discussing the tribulations of his patients, Rosenberg winces visibly; no one who has spent time with him could call him insensitive. He is, however, as psychiatrists say, well defended. "My goal is to find better treatments for people with cancer," he says. "I would never do something to a patient that I think would be hurtful. But there's a limit to what animal models can teach you."

Rosenberg was only a young boy when he first learned that suffering could lead to enlightenment; he saw his parents' generation sacrifice so that their children would not have to endure what they endured. And so it remains for him, at sixty. If there is pain in bearing witness to the dying, he has learned that this is the price. Evil, in his worldview, may have a purpose after all. Evil teaches.

Tonight, Rosenberg will leave the hospital for a pickup basketball game. It's one of the few activities that completely relaxes him — and the only thing he watches on TV. Before he leaves, however, he needs to do the one thing he's been both dreading and looking forward to all day: check his young patient's X rays to determine the effects of treatment on his tumors. He practically runs to the X-ray room.

The effects of the teenager's last treatment, administered six weeks ago, have been brutal. First, tumor-infiltrating lymphocytes that had been observed trying to attack his tumors were isolated, removed, and grown in great quantities in the lab. Next, with hideously aggressive chemotherapy, his immune system was almost obliterated, destroying virtually all the lymphocytes in his body. The theory in play here is that the body seems to contain a finite number of white blood cells of various sorts; it is hoped that, once the body is filled with sufficient lymphocytes, any extras that appear will just die off. So if you get rid of all the lymphocytes temporarily and then infuse the body with pure tumor-fighting cells, maybe, just maybe, this young man will become a tumor-fighting machine.

Rosenberg stares at the X rays. There is no question: the tumors have shrunk a lot. The tumor on the chest wall has clear enough margins that it might be possible to remove it completely. "Oh, boy, oh, boy," Rosenberg repeats to himself. "This is really exciting, because nothing has worked for this kid before."

That was November 2. On December 27, at the boy's next appointment, the tumors had continued to shrink. What, if anything, had made the difference this time? Nobody knew.

By late January the tumors had continued to shrink, enough that he was able to go to college.

"This is an astonishing case," Rosenberg says. "I've now cloned the gene that encodes the antigen. So we're studying this boy literally day and night. We've got to be careful — he's just one patient showing a response to this particular treatment. My heart's in my mouth to see what happens."

And so, in a sense, are all our hearts.

DENNIS OVERBYE

How Islam Won, and Lost, the Lead in Science

FROM *The New York Times*

NASIR AL-DIN AL-TUSI was still a young man when the Assassins made him an offer he couldn't refuse.

His hometown had been devastated by Mongol armies, and so, early in the thirteenth century, al-Tusi, a promising astronomer and philosopher, came to dwell in the legendary fortress city of Alamut in the mountains of northern Persia. He lived among a heretical and secretive sect of Shiite Muslims, whose members practiced political murder as a tactic and were dubbed *hashishinn,* legend has it, because of their use of hashish.

Although al-Tusi later said he had been held in Alamut against his will, the library there was renowned for its excellence, and al-Tusi thrived there, publishing works on astronomy, ethics, mathematics, and philosophy that marked him as one of the great intellectuals of his age. But when the armies of Halagu, the grandson of Genghis Khan, massed outside the city in 1256, al-Tusi had little trouble deciding where his loyalties lay. He joined Halagu and accompanied him to Baghdad, which fell in 1258. The grateful Halagu built him an observatory at Maragha, in what is now northwestern Iran.

Al-Tusi's deftness and ideological flexibility in pursuit of the resources to do science paid off. The road to modern astronomy, scholars say, leads through the work that he and his followers performed at Maragha and Alamut in the thirteenth and fourteenth centuries. It is a road that winds from Athens to Alexandria, Baghdad, Damascus, and Córdoba, through the palaces of caliphs and the basement laboratories of alchemists, and it was traveled not just by astronomy but by all science.

Commanded by the Koran to seek knowledge and read nature for signs of the Creator, and inspired by a treasure trove of ancient Greek learning, Muslims created a society that in the Middle Ages was the scientific center of the world. The Arabic language was synonymous with learning and science for five hundred years, a golden age that can count among its credits the precursors to modern universities, algebra, the names of the stars, and even the notion of science as an empirical inquiry. "Nothing in Europe could hold a candle to what was going on in the Islamic world until about 1600," said Dr. Jamil Ragep, a professor of the history of science at the University of Oklahoma.

It was the infusion of this knowledge into Western Europe, historians say, that fueled the Renaissance and the scientific revolution. "Civilizations don't just clash," said Dr. Abdelhamid Sabra, a retired professor of the history of Arabic science who taught at Harvard. "They can learn from each other. Islam is a good example of that." The intellectual meeting of Arabia and Greece was one of the greatest events in history, he said. "Its scale and consequences are enormous, not just for Islam but for Europe and the world."

But historians say they still know very little about this golden age. Few of the major scientific works from that era have been translated from Arabic, and thousands of manuscripts have never even been read by modern scholars. Dr. Sabra characterizes the history of Islamic science as a field that "hasn't even begun yet."

Islam's rich intellectual history, scholars are at pains and seem saddened and embarrassed to point out, belies the image cast by recent world events. Traditionally, Islam has encouraged science and learning. "There is no conflict between Islam and science," said Dr. Osman Bakar of the Center for Muslim-Christian Understanding at Georgetown. "Knowledge is part of the creed," added Dr. Farouk El-Baz, a geologist at Boston University, who was science adviser to President Anwar el-Sadat of Egypt. "When you know more, you see more evidence of God."

So the notion that modern Islamic science is now considered "abysmal," as Abdus Salam, the first Muslim to win a Nobel Prize in physics, once put it, haunts Eastern scholars. "Muslims have a kind of nostalgia for the past, when they could contend that they were the dominant cultivators of science," Dr. Bakar said. The relation between science and religion has generated much debate in the Islamic world, he and other scholars said. Some scientists and historians call for an "Islamic science" informed by spiritual values they say Western science ignores, but others argue that a religious conservatism in the East has dampened the skeptical spirit necessary for good science.

When Muhammad's armies swept out from the Arabian peninsula in the seventh and eighth centuries, annexing territory from Spain to Persia, they also annexed the works of Plato, Aristotle, Democritus, Pythagoras, Archimedes, Hippocrates, and other Greek thinkers. Hellenistic culture had been spread eastward by the armies of Alexander the Great and by religious minorities, including various Christian sects, according to Dr. David Lindberg, a medieval science historian at the University of Wisconsin. The largely illiterate Muslim conquerors turned to the local intelligentsia to help them govern, Dr. Lindberg said. In the process, he said, they absorbed Greek learning that had yet to be transmitted to the West in a serious way, or even translated into Latin. "The West had a thin version of Greek knowledge," Dr. Lindberg said. "The East had it all."

In ninth-century Baghdad, the caliph Abu al-Abbas al-Mamun set up an institute, the House of Wisdom, to translate manuscripts. Among the first works rendered into Arabic was the Alexandrian astronomer Ptolemy's *Great Work*, which described a universe in which the sun, moon, planets, and stars revolved around earth; *Al-Magest*, as the work was known to Arabic scholars, became the basis for cosmology for the next five hundred years.

Jews, Christians, and Muslims all participated in this flowering of science, art, medicine, and philosophy, which endured for at least five hundred years and spread from Spain to Persia. Its height, historians say, was in the tenth and eleventh centuries, when three great thinkers strode the East: Abu Ali al-Hasan ibn al-Haytham, also known as Alhazen; Abu Rayham Muhammad al-Biruni; and Abu Ali al-Hussein Ibn Sina, also known as Avicenna.

Al-Haytham, born in Iraq in 965, experimented with light and vision, laying the foundation for modern optics and for the notion that science should be based on experiment as well as on philosophical arguments. "He ranks with Archimedes, Kepler, and Newton as a great mathematical scientist," said Dr. Lindberg.

The mathematician, astronomer, and geographer al-Biruni, born in what is now part of Uzbekistan in 973, wrote some 146 works totaling 13,000 pages, including a vast sociological and geographical study of India.

Ibn Sina was a physician and philosopher born near Bukhara (now in Uzbekistan) in 981. He compiled a million-word medical encyclopedia, the Canons of Medicine, that was used as a textbook in parts of the West until the seventeenth century.

Scholars say science found such favor in medieval Islam for several reasons. Part of the allure was mystical; it was another way to experience the unity of creation that was the central message of Islam. "Anyone who studies anatomy will increase his faith in the omnipotence and oneness of God the Almighty," goes a saying often attributed to Abul-Walid Muhammad Ibn Rushd, also known as Averroës, a thirteenth-century anatomist and philosopher.

Another reason is that Islam is one of the few religions in human history in which scientific procedures are necessary for religious ritual, Dr. David King, a historian of science at Johann Wolfgang Goethe University in Frankfurt, pointed out in his book *Astronomy in the Service of Islam,* published in 1993. Arabs had always been knowledgeable about the stars and used them to navigate the desert, but Islam raised the stakes for astronomy.

The requirement that Muslims face in the direction of Mecca when they pray, for example, required knowledge of the size and shape of the earth. The best astronomical minds of the Muslim world tackled the job of producing tables or diagrams by which the qibla, or sacred directions, could be found from any point in the Islamic world. Their efforts rose to a precision far beyond the needs of the peasants who would use them, noted Dr. King. Astronomers at the Samarkand observatory, which was founded about 1420 by the ruler Ulugh Beg, measured star positions to a fraction of a degree, said Dr. El-Baz.

Islamic astronomy reached its zenith, at least from the Western perspective, in the thirteenth and fourteenth centuries, when al-Tusi and his successors pushed against the limits of the Ptolemaic worldview that had ruled for a millennium. According to the philosophers, celestial bodies were supposed to move in circles at uniform speeds. But the beauty of Ptolemy's attempt to explain the very un-uniform motions of planets and the sun as seen from earth was marred by corrections like orbits within orbits, known as epicycles, and geometrical modifications. Al-Tusi found a way to restore most of the symmetry to Ptolemy's model by adding pairs of cleverly designed epicycles to each orbit. Following in al-Tusi's footsteps, the fourteenth-century astronomer Ala al-Din Abul-Hasan ibn al-Shatir had managed to go further and construct a completely symmetrical model.

Copernicus, who overturned the Ptolemaic universe in 1530 by proposing that the planets revolved around the sun, expressed ideas similar to those of the Muslim astronomers in his early writings. This has led some historians to suggest that there is a previously unknown link between Copernicus and the Islamic astronomers, even though neither ibn al-Shatir's nor al-Tusi's work is known to have ever been translated into Latin, and therefore was presumably unknown in the West.

Dr. Owen Gingerich, an astronomer and historian of astronomy at Harvard, said he believed that Copernicus could have developed the ideas independently, but wrote in *Scientific American* that the whole idea of criticizing Ptolemy and reforming his model was part of "the climate of opinion inherited by the Latin West from Islam."

Despite their awareness of Ptolemy's flaws, Islamic astronomers were a long way from throwing out his model: dismissing it would have required a philosophical as well as cosmological revolution. "In some ways it was beginning to happen," said Dr. Ragep of the University of Oklahoma. But the East had no need of heliocentric models of the universe, said Dr. King of Frankfurt. All motion being relative, he said, it was irrelevant for the purposes of Muslim rituals whether the sun went around the earth or vice versa.

From the tenth to the thirteenth century, Europeans, especially in Spain, were translating Arabic works into Hebrew and Latin "as fast as they could," said Dr. King. The result was a rebirth of learning that ultimately transformed Western civilization.

Why didn't Eastern science go forward as well? "Nobody has answered that question satisfactorily," said Dr. Sabra of Harvard. Pressed, historians offer up a constellation of reasons. Among other things, the Islamic empire began to be whittled away in the thirteenth century by Crusaders from the West and Mongols from the East.

Christians reconquered Spain and its magnificent libraries in Córdoba and Toledo, full of Arab learning. As a result, Islamic centers of learning began to lose touch with one another and with the West, leading to a gradual erosion in two of the main pillars of science — communication and financial support.

In the West, science was able to pay for itself in new technology like the steam engine and to attract financing from industry, but in the East it remained dependent on the patronage and curiosity of sultans and caliphs. Further, the Ottomans, who took over the Arabic lands in the sixteenth century, were builders and conquerors, not thinkers, said Dr. El-Baz of Boston University, and support waned. "You cannot expect the science to be excellent while the society is not," he said.

Others argue, however, that Islamic science seems to decline only when viewed through Western, secular eyes. "It's possible to live without an industrial revolution if you have enough camels and food," Dr. King said. "Why did Muslim science decline?" he said. "That's a very Western question. It flourished for a thousand years — no civilization on earth has flourished that long in that way."

Humiliating encounters with Western colonial powers in the nineteenth century produced a hunger for Western science and technology, or at least the economic and military power they could produce, scholars say. Reformers bent on modernizing Eastern educational systems to include Western science could argue that Muslims would only be reclaiming their own, since the West had inherited science from the Islamic world to begin with.

In some ways these efforts have been very successful. "In particular countries the science syllabus is quite modern," said Dr. Bakar of Georgetown, citing Malaysia, Jordan, and Pakistan, in particular. Even in Saudi Arabia, one of the most conservative Muslim states, science classes are conducted in English, Dr. Sabra said.

Nevertheless, science still lags in the Muslim world, according to Dr. Pervez Hoodbhoy, a Pakistani physicist and professor at Quaid-e-Azam University in Islamabad, who has written on Islam and science. According to his own informal survey, included in his 1991 book *Islam and Science: Religious Orthodoxy and the Battle for Rational-*

ity, Muslims are seriously underrepresented in science, accounting for fewer than one percent of the world's scientists while they account for almost a fifth of the world's population. Israel, he reports, has almost twice as many scientists as the Muslim countries put together.

Among other sociological and economic factors, like the lack of a middle class, Dr. Hoodbhoy attributes the malaise of Muslim science to an increasing emphasis over the last millennium on rote learning based on the Koran. "The notion that all knowledge is in the Great Text is a great disincentive to learning," he said. "It's destructive if we want to create a thinking person, someone who can analyze, question, and create." Dr. Bruno Guideroni, a Muslim who is an astrophysicist at the National Center for Scientific Research in Paris, said, "The fundamentalists criticize science simply because it is Western."

Other scholars said the attitude of conservative Muslims to science was not so much hostile as schizophrenic, wanting its benefits but not its worldview. "They may use modern technology, but they don't deal with issues of religion and science." said Dr. Bakar. One response to the invasion of Western science, said the scientists, has been an effort to "Islamicize" science by portraying the Koran as a source of scientific knowledge.

Dr. Hoodbhoy said such groups had criticized the concept of cause and effect. Educational guidelines once issued by the Institute for Policy Studies in Pakistan, for example, included the recommendation that physical effects not be related to causes. For example, it was not Islamic to say that combining hydrogen and oxygen makes water. "You were supposed to say," Dr. Hoodbhoy recounted, "that when you bring hydrogen and oxygen together, then by the will of Allah water was created."

Even Muslims who reject fundamentalism, however, have expressed doubts about the desirability of following the Western style of science, saying that it subverts traditional spiritual values and promotes materialism and alienation. "No science is created in a vacuum," said Dr. Seyyed Hossein Nasr, a science historian, author, philosopher, and professor of Islamic studies at George Washington University, during a speech at the Massachusetts Institute of Technology a few years ago. "Science arose under particular circumstances in the West with certain philosophical presumptions about the nature of reality."

Dr. Muzaffar Iqbal, a chemist and the president and founder of the Center for Islam and Science in Alberta, Canada, explained: "Modern science doesn't claim to address the purpose of life; that is outside the domain. In the Islamic world, purpose is integral, part of that life."

Most working scientists tend to scoff at the notion that science can be divided into ethnic, religious, or any other kind of flavor. There is only one universe. The process of asking and answering questions about nature, they say, eventually erases the particular circumstances from which those questions arise.

In his book, Dr. Hoodbhoy recounts how Dr. Salam, Dr. Steven Weinberg, now at the University of Texas, and Dr. Sheldon Glashow at Harvard shared the Nobel Prize for showing that electromagnetism and the so-called weak nuclear force are different manifestations of a single force. Dr. Salam and Dr. Weinberg had devised the same contribution to that theory independently, he wrote, despite the fact that Dr. Weinberg is an atheist while Dr. Salam was a Muslim who prayed regularly and quoted from the Koran. Dr. Salam confirmed the account in his introduction to the book, describing himself as "geographically and ideologically remote" from Dr. Weinberg.

"Science is international," said Dr. El-Baz. "There is no such thing as Islamic science. Science is like building a big building, a pyramid. Each person puts up a block. These blocks have never had a religion. It's irrelevant, the color of the guy who put up the block."

CHET RAYMO

A Little Reminder of Reality's Scale

FROM *The Boston Globe*

I HAVE a biologist colleague who knows what a fellow likes. As a retirement gift, she gave me a bottle containing a few ounces of water, some algae, assorted microscopic organisms, and — wonder of wonders! — a few tardigrades.

She knew I would be appreciative. On a few occasions over the years, I had mentioned to her how much I would like to see a tardigrade in the flesh. These little creatures, about the size of the period at the end of this sentence, are adorably cute in microphotographs. And here they were, cavorting like playful otters in the field of view of my microscope. Tardigrades — literally, "slow walkers" — are sometimes called "water bears" because of the way they lumber along bearlike on eight stumpy appendages, or even more charmingly, "moss piglets." Under the microscope, they do indeed look remarkably like vertebrates of some sort, but they have no bony skeleton. They are invertebrates, related to insects, but so unique they have a phylum all of their own.

Tardigrades do not interest scientists just because they are cute. They are also among the hardiest of multicelled animals, maybe the toughest little critters of all. Dry them out and they go into a state of suspended animation in which they can live for — well, no one knows. When some apparently lifeless, 120-year-old moss from an Italian museum was moistened, tardigrades rose as if from the dead and scampered about. They can be frozen at temperatures near absolute zero, heated to 150 degrees centigrade, subjected to high vacuum or to pressure greater than that of the deepest ocean, and zapped with deadly radiation. It is not impossible that tardigrades could survive space travel without a spaceship.

Some scientists are trying to learn the tardigrade's secret of surviving cold in order to keep frozen human organs fresh and viable for transplants. My own curiosity, I confess, was based entirely on the tardigrade's reputation as a water bear or moss piglet. I mean, who can resist a creature the size of a dust-mote that looks like something out of Beatrix Potter?

Scale is the secret of the tardigrade's charm. It looks like it should be much bigger than it is, by about four orders of magnitude. It looks like a beast you might meet wallowing on a farm or tipping over trash cans in a national park. But you will more likely find them — if you have a magnifier — in wet moss in the gutters of your house.

For an hour, I observed my tardigrades scampering among strands of algae in a petri dish of water. They curled, stretched, crept, and reached, presumably grazing, although I never saw them feed, which they do by sucking juices out of microscopic prey.

Watching them, I became more conscious than ever of just how much we are prisoners of scale. Another whole universe exists down there on the microscopic scale, and below. Electron-microscope images of tardigrades show every pore and bristle, the little hook-shaped appendages that pass for toes, and long wavy "hairs" that look wildly unkempt.

It would be fun if we could shrink ourselves like Alice and frolic with the tardigrades. Then shrink further and swim with the ciliates and rotifers that buzzed about my tardigrades like bees, and further still to observe the amoebic creatures too small to see with my microscope.

But why stop with the smallest creatures? Keep shrinking, down to the level of molecules and atoms, and observe those misty clouds of electric charge that are the building blocks of the universe — God's Tinker Toys.

Even these are not the bottom floor. Smaller than the atoms are the quarks, elusive subatomic particles that hold themselves together in two's and three's so tightly that, so far, scientists haven't been able to pry them apart. We describe tardigrades with familiar metaphors — water bears or moss piglets — but quarks are so far beyond the scale of ordinary experience that all metaphors fail.

And many physicists believe that even quarks are not the ground floor of reality. They are searching for things called "strings," "quantum loops," or "spin foam" that exist on a scale twenty orders of magnitude smaller than the nucleus of an atom, at a level of reality where even space and time break up into their smallest, indivisible units.

As I peered into my microscope, I was thinking of this almost unimaginable shadow world on a scale 10 thousand billion billion billion times smaller than my water bears, that in its exotic shimmerings gives rise to the spectacular world of our senses.

ERIC SCHLOSSER

Why McDonald's Fries Taste So Good

FROM *The Atlantic Monthly*

THE FRENCH FRY was "almost sacrosanct for me," Ray Kroc, one of the founders of McDonald's, wrote in his autobiography, "its preparation a ritual to be followed religiously." During the chain's early years french fries were made from scratch every day. Russet Burbank potatoes were peeled, cut into shoestrings, and fried in McDonald's kitchens. As the chain expanded nationwide, in the mid-1960s, it sought to cut labor costs, reduce the number of suppliers, and ensure that its fries tasted the same at every restaurant. McDonald's began switching to frozen french fries in 1966 — and few customers noticed the difference. Nevertheless, the change had a profound effect on the nation's agriculture and diet. A familiar food had been transformed into a highly processed industrial commodity. McDonald's fries now come from huge manufacturing plants that can peel, slice, cook, and freeze 2 million pounds of potatoes a day. The rapid expansion of McDonald's and the popularity of its low-cost, mass-produced fries changed the way Americans eat. In 1960 Americans consumed an average of about eighty-one pounds of fresh potatoes and four pounds of frozen french fries. In 2000 they consumed an average of about fifty pounds of fresh potatoes and thirty pounds of frozen fries. Today McDonald's is the largest buyer of potatoes in the United States.

The taste of McDonald's french fries played a crucial role in the chain's success — fries are much more profitable than hamburgers — and was long praised by customers, competitors, and even food critics. James Beard loved McDonald's fries. Their distinctive taste does not stem from the kind of potatoes that McDonald's buys, the technology that processes them, or the restaurant equipment that fries them: other chains use Russet Burbanks, buy their french fries from the same large processing companies, and have similar fryers in their restaurant kitchens. The taste of a french fry is largely determined by the cooking oil. For decades McDonald's cooked its french fries in a mixture of about 7 percent cottonseed oil and 93 percent beef tallow. The mixture gave the fries their unique flavor — and more saturated beef fat per ounce than a McDonald's hamburger.

In 1990, amid a barrage of criticism over the amount of cholesterol in its fries, McDonald's switched to pure vegetable oil. This presented the company with a challenge: how to make fries that subtly taste like beef without cooking them in beef tallow. A look at the ingredients in McDonald's french fries suggests how the problem was solved. Toward the end of the list is a seemingly innocuous yet oddly mysterious phrase: "natural flavor." That ingredient helps to explain not only why the fries taste so good but also why most fast food — indeed, most of the food Americans eat today — tastes the way it does.

Open your refrigerator, your freezer, your kitchen cupboards, and look at the labels on your food. You'll find "natural flavor" or "artificial flavor" in just about every list of ingredients. The similarities between these two broad categories are far more significant than the differences. Both are man-made additives that give most processed food most of its taste. People usually buy a food item the first time because of its packaging or appearance. Taste usually determines whether they buy it again. About 90 percent of the money that Americans now spend on food goes to buy processed food. The canning, freezing, and dehydrating techniques used in processing destroy most of food's flavor — and so a vast industry has arisen in the United States to make processed food palatable. Without this flavor industry today's fast food would not exist. The names of the leading American fast-food chains and their best-selling menu items have become embedded in our popular culture and famous worldwide. But few people can name the companies that manufacture fast food's taste.

The flavor industry is highly secretive. Its leading companies will not divulge the precise formulas of flavor compounds or the identities of clients. The secrecy is deemed essential for protecting the reputations of beloved brands. The fast-food chains, understandably, would like the public to believe that the flavors of the food they sell somehow originate in their restaurant kitchens, not in distant factories run by other firms. A McDonald's french fry is one of countless foods whose flavor is just a component in a complex manufacturing process. The look and the taste of what we eat now are frequently deceiving — by design.

The New Jersey Turnpike runs through the heart of the flavor industry, an industrial corridor dotted with refineries and chemical plants. International Flavors & Fragrances (IFF), the world's largest flavor company, has a manufacturing facility off Exit 8A in Dayton, New Jersey; Givaudan, the world's second-largest flavor company, has a plant in East Hanover. Haarmann & Reimer, the largest German flavor company, has a plant in Teterboro, as does Takasago, the largest Japanese flavor company. Flavor Dynamics has a plant in South Plainfield; Frutarom is in North Bergen; Elan Chemical is in Newark. Dozens of companies manufacture flavors in the corridor between Teaneck and South Brunswick. Altogether the area produces about two-thirds of the flavor additives sold in the United States.

The IFF plant in Dayton is a huge pale-blue building with a modern office complex attached to the front. It sits in an industrial park, not far from a BASF plastics factory, a Jolly French Toast factory, and a plant that manufactures Liz Claiborne cosmetics. Dozens of tractor-trailers were parked at the IFF loading dock the afternoon I visited, and a thin cloud of steam floated from a roof vent. Before entering the plant, I signed a nondisclosure form, promising not to reveal the brand names of foods that contain IFF flavors. The place reminded me of Willy Wonka's chocolate factory. Wonderful smells drifted through the hallways, men and women in neat white lab coats cheerfully went about their work, and hundreds of little glass bottles sat on laboratory tables and shelves. The bottles contained powerful but fragile flavor chemicals, shielded from light by brown glass and round white caps shut tight. The long chemical names on the little white labels were as mystifying to me as medieval Latin. These odd-sounding things would be mixed and poured and turned into new substances, like magic potions.

I was not invited into the manufacturing areas of the IFF plant, where, it was thought, I might discover trade secrets. Instead I toured various laboratories and pilot kitchens, where the flavors of well-established brands are tested or adjusted, and where whole new flavors are created. IFF's snack-and-savory lab is responsible for the flavors of potato chips, corn chips, breads, crackers, breakfast cereals, and pet food. The confectionery lab devises flavors for ice cream, cookies, candies, toothpastes, mouthwashes, and antacids. Everywhere I looked, I saw famous, widely advertised products sitting on laboratory desks and tables. The beverage lab was full of brightly colored liquids in clear bottles. It comes up with flavors for popular soft drinks, sports drinks, bottled teas, and wine coolers, for all-natural juice drinks, organic soy drinks, beers, and malt liquors. In one pilot kitchen I saw a dapper food technologist, a middle-aged man with an elegant tie beneath his crisp lab coat, carefully preparing a batch of cookies with white frosting and pink-and-white sprinkles. In another pilot kitchen I saw a pizza oven, a grill, a milk-shake machine, and a french fryer identical to those I'd seen at innumerable fast-food restaurants.

In addition to being the world's largest flavor company, IFF manufactures the smells of six of the ten best-selling fine perfumes in the United States, including Estée Lauder's Beautiful, Clinique's Happy, Lancôme's Trésor, and Calvin Klein's Eternity. It also makes the smells of household products such as deodorant, dishwashing detergent, bath soap, shampoo, furniture polish, and floor wax. All these aromas are made through essentially the same process: the manipulation of volatile chemicals. The basic science behind the scent of your shaving cream is the same as that governing the flavor of your TV dinner.

Scientists now believe that human beings acquired the sense of taste as a way to avoid being poisoned. Edible plants generally taste sweet, harmful ones bitter. The taste buds on our tongues can detect the presence of half a dozen or so basic tastes, including sweet, sour, bitter, salty, astringent, and umami, a taste discovered by Japanese researchers — a rich and full sense of deliciousness triggered by amino acids in foods such as meat, shellfish, mushrooms, potatoes, and seaweed. Taste buds offer a limited means of detection, however, compared with the human olfactory system, which can perceive thousands of different chemical aromas. Indeed, "flavor" is primarily the smell of gases being released by the chemicals you've just put in your mouth. The aroma of a food can be responsible for as much as 90 percent of its taste.

The act of drinking, sucking, or chewing a substance releases its volatile gases. They flow out of your mouth and up your nostrils, or up the passageway in the back of your mouth, to a thin layer of nerve cells called the olfactory epithelium, located at the base of your nose, right between your eyes. Your brain combines the complex smell signals from your olfactory epithelium with the simple taste signals from your tongue, assigns a flavor to what's in your mouth, and decides if it's something you want to eat.

A person's food preferences, like his or her personality, are formed during the first few years of life, through a process of socialization. Babies innately prefer sweet tastes and reject bitter ones; toddlers can learn to enjoy hot and spicy food, bland health food, or fast food, depending on what the people around them eat. The human sense of smell is still not fully understood. It is greatly affected by psychological factors and expectations. The mind focuses intently on some of the aromas that surround us and filters out the overwhelming majority. People can grow accustomed to bad smells or good smells; they stop noticing what once seemed overpowering. Aroma and memory are somehow inextricably linked. A smell can suddenly evoke a long-forgotten moment. The flavors of childhood foods seem to leave an indelible mark, and adults often return to them, without always knowing why. These "comfort foods" become a source of pleasure and reassurance — a fact that fast-food chains use to their advantage. Childhood memories of Happy Meals, which come with french fries, can translate into frequent adult visits to McDonald's. On average, Americans now eat about four servings of french fries every week.

The human craving for flavor has been a largely unacknowledged and unexamined force in history. For millennia royal empires have been built, unexplored lands traversed, and great religions and philosophies forever changed by the spice trade. In 1492 Christopher Columbus set sail to find seasoning. Today the influence of flavor in the world marketplace is no less decisive. The rise and fall of corporate empires — of soft-drink companies, snack-food companies, and fast-food chains — is often determined by how their products taste.

The flavor industry emerged in the mid-nineteenth century, as processed foods began to be manufactured on a large scale. Recognizing the need for flavor additives, early food processors turned to perfume companies that had long experience working with essential oils and volatile aromas. The great perfume houses of England, France, and the Netherlands produced many of the first flavor compounds. In the early part of the twentieth century Germany took the technological lead in flavor production, owing to its powerful chemical industry. Legend has it that a German scientist discovered methyl anthranilate, one of the first artificial flavors, by accident while mixing chemicals in his laboratory. Suddenly the lab was filled with the sweet smell of grapes. Methyl anthranilate later became the chief flavor compound in grape Kool-Aid. After World War II much of the perfume industry shifted from Europe to the United States, settling in New York City near the garment district and the fashion houses. The flavor industry came with it, later moving to New Jersey for greater plant capacity. Man-made flavor additives were used mostly in baked goods, candies, and sodas until the 1950s, when sales of processed food began to soar. The invention of gas chromatographs and mass spectrometers — machines capable of detecting volatile gases at low levels — vastly increased the number of flavors that could be synthesized. By the mid-1960s flavor companies were churning out compounds to supply the taste of Pop Tarts, Bac-Os, Tab, Tang, Filet-O-Fish sandwiches, and literally thousands of other new foods.

The American flavor industry now has annual revenues of about $1.4 billion. Approximately ten thousand new processed-food products are introduced every year in the United States. Almost all of them require flavor additives. And about nine out of ten of these products fail. The latest flavor innovations and corporate realignments are heralded in publications such as *Chemical Market Reporter, Food Chemical News, Food Engineering,* and *Food Product Design.* The progress of IFF has mirrored that of the flavor industry as a whole. IFF was formed in 1958, through the merger of two small companies. Its annual revenues have grown almost fifteenfold since the early 1970s, and it currently has manufacturing facilities in twenty countries.

Today's sophisticated spectrometers, gas chromatographs, and headspace-vapor analyzers provide a detailed map of a food's flavor components, detecting chemical aromas present in amounts as low as one part per billion. The human nose, however, is even more sensitive. A nose can detect aromas present in quantities of a few parts per trillion — an amount equivalent to about 0.0000000000003 percent. Complex aromas, such as those of coffee and roasted meat, are composed of volatile gases from nearly a thousand different chemicals. The smell of a strawberry arises from the interaction of about 350 chemicals that are present in minute amounts. The quality that people seek most of all in a food — flavor — is usually present in a quantity too infinitesimal to be measured in traditional culinary terms such as ounces or teaspoons. The chemical that provides the dominant flavor of bell pepper can be tasted in amounts as low as 0.02 parts per billion; one drop is sufficient to add flavor to five average-size swimming pools. The flavor additive usually comes next to last in a processed food's list of ingredients and often costs less than its packaging. Soft drinks contain a larger proportion of flavor additives than most products. The flavor in a twelve-ounce can of Coke costs about half a cent.

The color additives in processed foods are usually present in even smaller amounts than the flavor compounds. Many of New Jersey's flavor companies also manufacture these color additives, which are used to make processed foods look fresh and appealing. Food coloring serves many of the same decorative purposes as lipstick, eye shadow, mascara — and is often made from the same pigments. Titanium dioxide, for example, has proved to be an especially versatile mineral. It gives many processed candies, frostings, and icings their bright white color; it is a common ingredient in women's cosmetics; and it is the pigment used in many white oil paints and house paints. At Burger King, Wendy's, and McDonald's coloring agents have been added to many of the soft drinks, salad dressings, cookies, condiments, chicken dishes, and sandwich buns.

Studies have found that the color of a food can greatly affect how its taste is perceived. Brightly colored foods frequently seem to taste better than bland-looking foods, even when the flavor compounds are identical. Foods that somehow look off-color often seem to have off tastes. For thousands of years human beings have relied on visual cues to help determine what is edible. The color of fruit suggests whether it is ripe, the color of meat whether it is rancid. Flavor researchers sometimes use colored lights to modify the influence of visual cues during taste tests. During one experiment in the early 1970s people were served an oddly tinted meal of steak and french fries that appeared normal beneath colored lights. Everyone thought the meal tasted fine until the lighting was changed. Once it became apparent that the steak was actually blue and the fries were green, some people became ill.

The federal Food and Drug Administration does not require companies to disclose the ingredients of their color or flavor additives so long as all the chemicals in them are considered by the agency to be GRAS ("generally recognized as safe"). This enables companies to maintain the secrecy of their formulas. It also hides the fact that flavor compounds often contain more ingredients than the foods to which they give taste. The phrase "artificial strawberry flavor" gives little hint of the chemical wizardry and manufacturing skill that can make a highly processed food taste like strawberries.

A typical artificial strawberry flavor, like the kind found in a Burger King strawberry milk shake, contains the following ingredients: amyl acetate, amyl butyrate, amyl valerate, anethol, anisyl formate, benzyl acetate, benzyl isobutyrate, butyric acid, cinnamyl isobutyrate, cinnamyl valerate, cognac essential oil, diacetyl, dipropyl ketone, ethyl acetate, ethyl amyl ketone, ethyl butyrate, ethyl cinnamate, ethyl heptanoate, ethyl heptylate, ethyl lactate, ethyl methylphenylglycidate, ethyl nitrate, ethyl propionate, ethyl valerate, heliotropin, hydroxyphenyl-2-butanone (10 percent solution in alcohol), α-ionone, isobutyl anthranilate, isobutyl butyrate, lemon essential oil, maltol, 4-methylacetophenone, methyl anthranilate, methyl benzoate, methyl cinnamate, methyl heptine carbonate, methyl naphthyl ketone, methyl salicylate, mint essential oil, neroli essential oil, nerolin, neryl isobutyrate, orris butter, phenethyl alcohol, rose, rum ether, γ-undecalactone, vanillin, and solvent.

Although flavors usually arise from a mixture of many different volatile chemicals, often a single compound supplies the dominant aroma. Smelled alone, that chemical provides an unmistakable sense of the food. Ethyl-2-methyl butyrate, for example, smells just like an apple. Many of today's highly processed foods offer a blank palette: whatever chemicals are added to them will give them specific tastes. Adding methyl-2-pyridyl ketone makes something taste like popcorn. Adding ethyl-3-hydroxy butanoate makes it taste like marshmallow. The possibilities are now almost limitless. Without affecting appearance or nutritional value, processed foods could be made with aroma chemicals such as hexanal (the smell of freshly cut grass) or 3-methyl butanoic acid (the smell of body odor).

The 1960s were the heyday of artificial flavors in the United States. The synthetic versions of flavor compounds were not subtle, but they did not have to be, given the nature of most processed food. For the past twenty years food processors have tried hard to use only "natural flavors" in their products. According to the FDA, these must be derived entirely from natural sources — from herbs, spices, fruits, vegetables, beef, chicken, yeast, bark, roots, and so forth. Consumers prefer to see natural flavors on a label, out of a belief that they are more healthful. Distinctions between artificial and natural flavors can be arbitrary and somewhat absurd, based more on how the flavor has been made than on what it actually contains.

"A natural flavor," says Terry Acree, a professor of food science at Cornell University, "is a flavor that's been derived with an out-of-date technology." Natural flavors and artificial flavors sometimes contain exactly the same chemicals, produced through different methods. Amyl acetate, for example, provides the dominant note of banana flavor. When it is distilled from bananas with a solvent, amyl acetate is a natural flavor. When it is produced by mixing vinegar with amyl alcohol and adding sulfuric acid as a catalyst, amyl acetate is an artificial flavor. Either way it smells and tastes the same. "Natural flavor" is now listed among the ingredients of everything from Health Valley Blueberry Granola Bars to Taco Bell Hot Taco Sauce.

A natural flavor is not necessarily more healthful or purer than an artificial one. When almond flavor — benzaldehyde — is derived from natural sources, such as peach and apricot pits, it contains traces of hydrogen cyanide, a deadly poison. Benzaldehyde derived by mixing oil of clove and amyl acetate does not contain any cyanide. Nevertheless, it is legally considered an artificial flavor and sells at a much lower price. Natural and artificial flavors are now manufactured at the same chemical plants, places that few people would associate with Mother Nature.

The small and elite group of scientists who create most of the flavor in most of the food now consumed in the United States are called "flavorists." They draw on a number of disciplines in their work: biology, psychology, physiology, and organic chemistry. A flavorist is a chemist with a trained nose and a poetic sensibility. Flavors are created by blending scores of different chemicals in tiny amounts — a process governed by scientific principles but demanding a fair amount of art. In an age when delicate aromas and microwave ovens do not easily coexist, the job of the flavorist is to conjure illusions about processed food and, in the words of one flavor company's literature, to ensure "consumer likeability." The flavorists with whom I spoke were discreet, in keeping with the dictates of their trade. They were also charming, cosmopolitan, and ironic. They not only enjoyed fine wine but could identify the chemicals that give each grape its unique aroma. One flavorist compared his work to composing music. A well-made flavor compound will have a "top note" that is often followed by a "dry-down" and a "leveling-off," with different chemicals responsible for each stage. The taste of a food can be radically altered by minute changes in the flavoring combination. "A little odor goes a long way," one flavorist told me.

In order to give a processed food a taste that consumers will find appealing, a flavorist must always consider the food's "mouthfeel" — the unique combination of textures and chemical interactions that affect how the flavor is perceived. Mouthfeel can be adjusted through the use of various fats, gums, starches, emulsifiers, and stabilizers. The aroma chemicals in a food can be precisely analyzed, but the elements that make up mouthfeel are much harder to measure. How does one quantify a pretzel's hardness, a french fry's crispness? Food technologists are now conducting basic research in rheology, the branch of physics that examines the flow and deformation of materials. A number of companies sell sophisticated devices that attempt to measure mouthfeel. The TA.XT2i Texture Analyzer, produced by the Texture Technologies Corporation, of Scarsdale, New York, performs calculations based on data derived from as many as 250 separate probes. It is essentially a mechanical mouth. It gauges the most important rheological properties of a food — bounce, creep, breaking point, density, crunchiness, chewiness, gumminess, lumpiness, rubberiness, springiness, slipperiness, smoothness, softness, wetness, juiciness, spreadability, springback, and tackiness.

Some of the most important advances in flavor manufacturing are now occurring in the field of biotechnology. Complex flavors are being made using enzyme reactions, fermentation, and fungal and tissue cultures. All the flavors created by these methods — including the ones being synthesized by fungi — are considered natural flavors by the FDA. The new enzyme-based processes are responsible for extremely true-to-life dairy flavors. One company now offers not just butter flavor but also fresh creamy butter, cheesy butter, milky butter, savory melted butter, and super-concentrated butter flavor, in liquid or powder form. The development of new fermentation techniques, along with new techniques for heating mixtures of sugar and amino acids, have led to the creation of much more realistic meat flavors.

The McDonald's Corporation most likely drew on these advances when it eliminated beef tallow from its french fries. The company will not reveal the exact origin of the natural flavor added to its fries. In response to inquiries from *Vegetarian Journal,* however, McDonald's did acknowledge that its fries derive some of their characteristic flavor from "an animal source." Beef is the probable source, although other meats cannot be ruled out. In France, for example, fries are sometimes cooked in duck fat or horse tallow.

Other popular fast foods derive their flavor from unexpected ingredients. McDonald's Chicken McNuggets contain beef extracts, as does Wendy's Grilled Chicken Sandwich. Burger King's BK Broiler Chicken Breast Patty contains "natural smoke flavor." A firm called Red Arrow Products specializes in smoke flavor, which is added to barbecue sauces, snack foods, and processed meats. Red Arrow manufactures natural smoke flavor by charring sawdust and capturing the aroma chemicals released into the air. The smoke is captured in water and then bottled, so that other companies can sell food that seems to have been cooked over a fire.

The Vegetarian Legal Action Network recently petitioned the FDA to issue new labeling requirements for foods that contain natural flavors. The group wants food processors to list the basic origins of their flavors on their labels. At the moment vegetarians often have no way of knowing whether a flavor additive contains beef, pork, poultry, or shellfish. One of the most widely used color additives — whose presence is often hidden by the phrase "color added" — violates a number of religious dietary restrictions, may cause allergic reactions in susceptible people, and comes from an unusual source. Cochineal extract (also known as carmine or carminic acid) is made from the desiccated bodies of female *Dactylopius coccus Costa,* a small insect harvested mainly in Peru and the Canary Islands. The bug feeds on red cactus berries, and color from the berries accumulates in the females and their unhatched larvae. The insects are collected, dried, and ground into a pigment. It takes about seventy thousand of them to produce a pound of carmine, which is used to make processed foods look pink, red, or purple. Dannon strawberry yogurt gets its color from carmine, and so do many frozen fruit bars, candies, and fruit fillings, and Ocean Spray pink-grapefruit juice drink.

In a meeting room at IFF, Brian Grainger let me sample some of the company's flavors. It was an unusual taste test — there was no food to taste. Grainger is a senior flavorist at IFF, a soft-spoken chemist with graying hair, an English accent, and a fondness for understatement. He could easily be mistaken for a British diplomat or the owner of a West End brasserie with two Michelin stars. Like many in the flavor industry, he has an Old World, old-fashioned sensibility. When I suggested that IFF's policy of secrecy and discretion was out of step with our mass-marketing, brand-conscious, self-promoting age, and that the company should put its own logo on the countless products that bear its flavors, instead of allowing other companies to enjoy the consumer loyalty and affection inspired by those flavors, Grainger politely disagreed, assuring me that such a thing would never be done. In the absence of public credit or acclaim, the small and secretive fraternity of flavor chemists praise one another's work. By analyzing the flavor formula of a product, Grainger can often tell which of his counterparts at a rival firm devised it. Whenever he walks down a supermarket aisle, he takes a quiet pleasure in seeing the well-known foods that contain his flavors.

Grainger had brought a dozen small glass bottles from the lab. After he opened each bottle, I dipped a fragrance-testing filter into it — a long white strip of paper designed to absorb aroma chemicals without producing off notes. Before placing each strip of paper in front of my nose, I closed my eyes. Then I inhaled deeply, and one food after another was conjured from the glass bottles. I smelled fresh cherries, black olives, sautéed onions, and shrimp. Grainger's most remarkable creation took me by surprise. After closing my eyes, I suddenly smelled a grilled hamburger. The aroma was uncanny, almost miraculous — as if someone in the room were flipping burgers on a hot grill. But when I opened my eyes, I saw just a narrow strip of white paper and a flavorist with a grin.

DANIEL SMITH

Shock and Disbelief

FROM *The Atlantic Monthly*

ON THE COVER of a pamphlet I was sent recently appears a photograph of an elderly man with bright bolts of electricity shooting outward from his temples. His teeth are clenched. His eyes are squeezed shut. His hair is standing on end. Holding the man's head secure is a leather strap that resembles the restraint on a prisoner in the electric chair.

This is electroconvulsive therapy (ECT) — the psychiatric use of an electric current to stimulate a grand mal seizure — as seen through the eyes of the Citizens Commission on Human Rights, a lobbying group founded by the Church of Scientology and the most active and well-organized anti-ECT group in existence. It is a grim view, invoking coercion, barbarity, anguish — everything negative that has ever been associated with psychiatry. It is also the common view.

Last fall I saw a patient receive ECT at McLean Hospital, a private psychiatric facility in Belmont, Massachusetts. There, in a well-lit treatment room, attended by a nurse, a psychiatrist, and an anesthesiologist, a middle-aged man suffering from hallucinations and depression lay unconscious on his back while two electrode paddles were placed on his head. A button was pressed, and the patient's right foot twitched lightly. Shortly afterward the patient awoke and was given a snack before being escorted back to his room.

The contrast between image and reality is surprising. The procedure I saw at McLean reflects the way ECT has been administered for years, as cautiously and as formally as any other medical procedure — perhaps even more so, because of the awareness psychiatrists have of ECT's reputation as savage. Yet the popular image of ECT has persisted, sustained almost single-handedly, it sometimes seems, by the 1975 movie *One Flew Over the Cuckoo's Nest,* the release of which coincided with a decline in the use of ECT. In 1980 less than 3 percent of all psychiatric inpatients were being treated with the procedure, and by 1983, thirty-three states were in some way regulating it.

Although the public seemed willing to let ECT fall into obsolescence, many psychiatrists felt that they were losing a valuable and irreplaceable treatment. In 1985 the National Institutes of Health, in Bethesda, Maryland, called a three-day conference on electroconvulsive therapy. The first day of the conference passed without incident, as experts delivered lectures. On the second day, however, during an open discussion period, anger erupted on the floor of the conference hall. Former patients and even a few clinicians began protesting loudly. One of those present was Max Fink, then a professor of psychiatry at the State University of New York at Stony Brook and a pioneer in modern ECT research. As Fink remembers it, "They were shouting, 'How dare you even consider electroshock as a possibility! It has no place in the world! Everybody who does electroshock should be in jail!'"

When the conference resumed, a panel of "nonadvocate" experts forged a consensus statement in which they observed, with standoffish delicacy,

> Electroconvulsive therapy is the most controversial treatment in psychiatry. The nature of the treatment itself, its history of abuse, unfavorable media presentations, compelling testimony of former patients, special attention by the legal system, uneven distribution of ECT use among practitioners and facilities, and uneven access by patients all contribute to the controversial context in which the consensus panel has approached its task.

Today ECT has strengthened its position in the profession. Many psychiatrists, whether or not they actively administer the treatment, have come to appreciate its ability to ameliorate a range of mental illnesses, from depression to some forms of schizophrenia and catatonia. A 1993 commentary in the *New England Journal of Medicine* stated, "Electroconvulsive therapy is more firmly established than ever as an important method of treating certain severe forms of depression." The first phase of a National Institute of Mental Health–supported study, to be published this spring, found that ECT produced a greater than 95 percent remission rate in psychotically depressed patients — vastly higher than the rate for any drug on the market. When I talked with Fink recently, he told me, "ECT is the most effective antidepressant, antipsychotic, anticatatonic we have today." Other psychiatrists have been even more enthusiastic. One, T. George Bidder, has written that ECT is "one of the most effective treatments in all of medicine — with a therapeutic efficacy, in properly selected cases, comparable to some of the most potent and specific treatments available, such as penicillin in pneumonococcal pneumonia." Such endorsements have led to what looks like a renaissance for ECT: it is estimated that 100,000 patients are treated with it each year — nearly triple the number cited for 1980 by the NIMH.

Yet the attacks on the treatment are as virulent as ever. Activists continue to push for prohibitive legislation. In 1997 a bill that would effectively have made administering ECT a criminal act, punishable by a fine of up to $10,000 and/or up to six months in jail, was narrowly defeated in Texas. ECT has virtually disappeared from state-run psychiatric facilities, owing in large part to government regulation. To be treated, patients must almost always gain access to a private or academic hospital. This means that ECT is very rarely an option for poor patients — those without adequate insurance or access to information, or without the means to travel, for example, to a distant, well-equipped university hospital. A 1995 article in the *American Journal of Psychiatry* found that ECT was unavailable in more than a third of the 317 metropolitan areas nationwide that it surveyed. "The situation has reversed itself from where it was decades ago," says Richard Weiner, a professor of psychiatry at Duke University and the head of the American Psychiatric Association's Committee on ECT. "Many ECT patients used to be asylum patients. Now it's very hard to get ECT in such places, and its use has shifted to general hospitals and private psychiatric hospitals."

The stigma attached to ECT is in some ways a holdover from less scrupulous days of psychiatry. But one of the main reasons many people still consider ECT to be archaic and even destructive is that it continues to be painted as such by an unlikely trio of activist groups: a handful of former ECT patients, some dissenting psychiatrists, and the Church of Scientology. These groups have agitated for the complete elimination of ECT. They have pushed legislative attempts to limit or ban ECT. They have initiated and supported lawsuits against psychiatrists, hospitals, and ECT-device manufacturers. They claim that ECT is authoritarian, violent, and representative of everything that is wrong with the profession of psychiatry. And despite all medical evidence to the contrary, people are listening to them.

Electroconvulsive therapy emerged during a bleak period for psychiatry. In the first third of the twentieth century not much could be done for the mentally ill. Psychoanalysis, the dominant method of treatment, proved helpful to some wealthy patients complaining of the so-called minor illnesses: melancholy and neurosis. But it didn't do much for patients with more systemic afflictions, such as schizophrenia and manic-depressive illness. These patients were merely warehoused in vast state asylums, where conditions were appalling. Patients were abused, shackled, even surgically sterilized. Psychiatry's job seemed to be no more than brutal custodianship; psychiatrists could do no more than hope that their patients would recover spontaneously from their illnesses. Under these desperate circumstances some psychiatrists began experimenting with radical treatments: insulin coma, transorbital lobotomy, malarial fever.

One of these "somatic therapies" — Metrazol shock — seemed particularly promising, given the theory (now known to be untrue) that a "biological antagonism" existed between epilepsy and schizophrenia. A schizophrenic patient was injected with Metrazol, a drug similar to camphor. After a few minutes the patient would undergo a full-blown seizure: all the muscles in his body would convulse violently, his back would arch, his limbs would flail, his breathing would become shallow. Often he would vomit. It was a gruesome ordeal. The historian Edward Shorter, in *A History of Psychiatry* (1997), reported that a Swiss psychiatrist stopped using the treatment because it caused "agonizing fears of dying and crumbling away," and that a British doctor spoke of "the unseemly and tragic farce of an unwilling patient being pursued by a posse of nurses with me, a fully charged syringe in my hand, bringing up the rear." And yet, strangely, Metrazol shock worked pretty well. "Convulsive therapy," as it came to be called, opened wide vistas of possibility.

But no one really understood why inducing seizures made patients better. Even today there are only educated guesses. Some subscribe to the neuroendocrine hypothesis, which states that seizures cause a shift in the body's hormonal system. Others subscribe to what has been called the anticonvulsant view, which holds that, paradoxically, the whole purpose of causing a seizure is to tap into the brain's ability to stop that seizure naturally. In other words, the brain's anticonvulsant mechanism may alter the brain's neurochemistry, acting as a built-in antidepressant. Still others believe that it is the seizures themselves that change the level of chemicals in the brain. In 1990 a group of articles in the journal *Neuropsychopharmacology* examined all three possibilities without drawing any conclusions.

Regardless, from the beginning convulsive therapy proved promising. Ugo Cerletti, in the 1930s the chief of the Clinic for Nervous and Mental Diseases at the University of Rome, was among those who were impressed. But he considered that electricity might cause seizures more quickly, and thus in a less harrowing manner, than Metrazol. Earlier Cerletti had tested the neurological effects of electricity by conducting experiments on dogs. His first attempts were inauspicious: because he put one electrode in the dog's mouth and one in its anus, the bulk of the current passed through the dog's heart; half the dogs died of cardiac arrest. Lucio Bini, one of Cerletti's assistants, solved this problem by transferring the electrodes to the dogs' temples. Cerletti and his staff worked tirelessly, experimenting on animals that were brought to them each week by dog catchers. The results supported their hopes: it seemed that using electricity was an effective way to produce an epileptic fit. Before applying it to a human being, Cerletti's assistants visited a Rome slaughterhouse to observe an electrical device that was being used to incapacitate pigs prior to slaughter. They discovered that there was a wide margin between the amount of electricity that would create a seizure and the amount that would kill.

In the spring of 1938 "electroshock," as Cerletti called it, was ready to be tested on a human being. The subject was a Milanese man the Roman police had found wandering in the train station without a ticket, mumbling gibberish to himself. Shorter described the inaugural treatment.

> The patient, his head shaved, seemed quite indifferent to what was going on. A nurse placed the electrodes on his temples while an orderly put a rubber tube between his teeth to prevent him from biting his tongue . . . There was a crack of electricity. The patient's muscles jolted once . . .
>
> "Let's step it up to 90," said Cerletti.
>
> Another electrical crack. Another spasm. The patient lay motionless for a minute, then began to sing.
>
> "We'll try it one last time at a higher voltage," said Cerletti, "*poi basta* [and then enough]."
>
> At this point, the patient said, in a perfectly calm and reasonable voice, as though answering an exam question, "Look out! The first is pestiferous, the second mortiferous." The residents looked at each other puzzled.

Despite the primitive application, the patient responded quite well. He had ten more treatments and was released, "in good condition and well-oriented." After a year he had not relapsed significantly. This was no small feat; no one could remember any experiment that had shown nearly such promising results. Thereafter ECT spread quickly to European hospitals. By 1940 it had appeared in the United States. Psychiatrists were enthusiastic. One, whom Shorter quoted, wrote in the *British Journal of Psychiatry*, "Without ECT I would not have lasted out in psychiatry, as I would not have been able to tolerate the sadness and hopelessness of most mental illnesses."

ECT was a great step up. Patients did not vomit, as they did in the course of Metrazol shock, and they did not experience as much psychological trauma. But they did still have to suffer the effects of muscular convulsions, which were frequently excruciating, and which have contributed to the persistent image of ECT as a brutal form of treatment. Thrashing around on the treatment table, many patients bit their tongues and cheeks. Many suffered broken bones or serious spinal injuries. Sometimes a gang of orderlies and nurses was needed to prevent the patient from tossing himself off the table altogether. In addition, patients suffered memory loss. They would awake confused, unsure of where they were or what had happened, often forgetting events of the preceding weeks or months.

ECT was also drastically overused. Doctors in some hospitals would treat dozens of patients in one giant room, wheeling the device on a cart from bed to bed; patients were forced to watch the ordeal of those who came before them. One doctor in England treated some of his patients more than a thousand times each. In the 1950s Ewen Cameron, a psychiatrist at McGill University, in Montreal, "depatterned" his patients by giving them twelve treatments daily. Milledgeville State Hospital, in Georgia, for a time the largest asylum in the United States, had perhaps the worst history of abuse: it used what was known as the Georgia Power Cocktail to punish uncooperative patients.

The publicized experiences of famous patients treated privately with ECT bolstered the evidence against the treatment. The poet Sylvia Plath was subjected to ECT and wrote about it in her autobiographical novel *The Bell Jar:* "Then something bent down and took hold of me and shook me like the end of the world. Whee-ee-ee-ee-ee, it shrilled, through an air crackling with blue light, and with each flash a great jolt drubbed me till I thought my bones would break and the sap fly out of me like a split plant." (Later in the novel the narrator had a less unpleasant ECT experience.) Ernest Hemingway underwent a course of ECT at the Mayo Clinic, in Rochester, Minnesota, and wrote to his biographer, A. E. Hotchner: "What is the sense of ruining my head and erasing my memory, which is my capital, and putting me out of business? It was a brilliant cure, but we lost the patient." Soon afterward Hemingway shot himself. In 1972 Senator Thomas Eagleton had to withdraw as the presidential candidate George McGovern's running mate after it was revealed that he had been treated with ECT. And, of course, *One Flew Over the Cuckoo's Nest* bundled all the public's negative associations into the disturbing image of Jack Nicholson, mocking and playful one moment, writhing on a table the next, and finally catatonic — the result, in actuality, not of the ECT he received but of an off-camera lobotomy.

ECT all but disappeared in the 1970s, eclipsed by psychiatric drugs, which brought about, as Shorter called it, the "triumph of the biological." More and more drugs came on the market, offering a sophisticated biochemical arsenal for treating mental illness. In the 1980s, owing to more advanced neuro-imaging techniques, physiological sources were found for schizophrenia and manic-depressive illness. As is by now well known, psychiatry and neurology edged toward a permanent intimacy. Electroconvulsive therapy seemed more than a little outmoded.

But drugs have not been the complete answer to mental illness. They were and still are a frustratingly inexact method of treatment — with a long wait between the first pill and any sign of relief. Often they don't work at all. This can be fatal for a patient who is suicidally depressed. Moreover, some patients prove resistant to medication.

The psychiatric community set out to modernize ECT and improve its image. Researchers worked with manufacturers to modernize ECT devices, outfitting them with equipment to monitor heart rate and brain activity and upgrading the electricity used. The 1985 NIH conference was followed by a 1990 report by the American Psychiatric Association committee charged with introducing better standards for treatment. The problem of physical injuries had been solved by the administration of fast-acting anesthesia and muscle relaxants, which confine the effects of a seizure to the brain. Clinicians implemented an informed-consent procedure that detailed every aspect of ECT along with its benefits and risks — including the (slim) possibility of death. (According to the most recent report of the APA Committee on ECT, published this year, one death occurs for every eighty thousand treatments.) ECT became safer and more exact, and psychiatrists used it more selectively. Today ECT is frequently used to treat the elderly, a population highly susceptible to mental illness and sensitive to the side effects of medication. Because drugs can cause birth defects, ECT is also often the preferred psychiatric treatment for women during the early stages of pregnancy.

Some side effects do remain. Memory loss is the most prevalent and is the primary reason that ECT is not used more often. Patients may have gaps in their memory affecting several months preceding treatment, and may also have trouble "laying down" new memories for a couple of weeks following treatment. In a small number of patients these problems can persist for a much longer period of time. But to some people, the fact that ECT causes any memory loss at all is unacceptable. ECT's detractors focus their objections on this side effect.

*

Linda Andre, a tall, attractive woman in her early forties, is the director of the Committee for Truth in Psychiatry, a loose-knit organization of five hundred former ECT patients. I was directed to CTIP by Max Fink, who has had numerous run-ins with Andre. At a talk Fink gave some years ago in New York, Andre stood up in the audience and loudly protested his association with Somatics, one of the two largest U.S. manufacturers of ECT devices. (Fink says he has no financial links with any ECT-device manufacturer.) Andre has been to many psychiatric conferences. She is hardly ever afforded official time to speak. More often she simply rises from the crowd.

When Andre and I met recently, I mentioned Fink, and she shook her head. "Ah, Max Fink, my dear friend. Oh, that man. That man. Not an honest and ethical individual, shall we say? I cannot believe that the scientific press lets the stuff he says get through. I'm sure he told you that no one ever had memory loss from ECT, except maybe around the time of ECT itself, and that they don't want to remember. He probably told you that we're just exaggerating. And everybody has some memory loss. He keeps these positions because he can. Shock is his baby. He's been associated with it longer than anyone." Fink, it became clear, represents for Andre the epitome of psychiatric deception. In an unpublished article on ECT, Andre has written,

> After fifty years of giving electroshock, I can't believe Fink knows any less about the extent of permanent memory loss and disability than I do. I believe he and his fellow apologists are making a value judgment about the worth of their patients' memories and lives, and deciding on that basis to essentially trade brain damage for temporary relief of depression.

Andre has been the director of CTIP since 1992. She told me she first became involved in the organization in 1985, several months after she received fifteen "shock" treatments at the Payne Whitney Psychiatric Clinic, in New York City. Andre takes exception to the term "ECT," dismissing "electroconvulsive therapy" as "the elegant new PR-conscious name for 'electroshock.'" She says that she doesn't remember anything about her treatments, and that she was committed to the clinic against her will.

"Everything I know about getting electroshock is what I've been told," she says. "I don't remember anything about it. From what I understand, my brother basically tricked me into going into the hospital at a time when I was going through a lot of problems and had become a pain in the ass to him." Andre says she escaped from the hospital several times before her treatment began, and that each time her brother recommitted her. When she was finally released, she says, she had both retrograde and anterograde amnesia: she couldn't remember much of the previous four years, and she had difficulty creating memories of new events. One day, at home, she heard a woman named Marilyn Rice talking on the radio about ECT.

Rice is something of a legend in the world of ECT. In 1974 the distinguished medical writer Berton Roueché published an article about her in *The New Yorker,* disguising Rice, who had received ECT to treat a serious bout of depression, as "Natalie Parker." The article, titled "As Empty as Eve," depicted Rice's experience as a nightmarish erasure of memory. "There is a harrowing sense of confusion," Roueché wrote, "and then a full awakening in the midnight dark of total amnesia." Her sense of purpose bolstered by the article, Rice formed CTIP and began accumulating documentation that ECT causes, as she put it, "psychiatrically induced brain damage." She wrote letters to psychiatrists, government officials, newspapers and magazines, and other potential allies, and created a small network of ECT "survivors," as she called them. To one doctor she wrote, "I could easily set up a psychiatric hospital as good as yours. I would just put the patients down on the sidewalk and interfere with their cerebral function by dropping flower pots on their heads."

Andre got Rice's phone number from the radio station and called her. The two became very close, and when Rice died, of heart failure, in 1992, Andre took over as director.

Can ECT cause a complete erasure of memory, as Andre claims? Most psychiatrists insist that it can't, and that side effects are usually slight. Roland Kohloff, the principal tympanist for the New York Philharmonic, was treated several times with ECT after slipping into severe depression. Each time he quickly rebounded and went back to work. "After you get a series," he told me recently, "there will be for a while some short-term-memory problems. I might not remember something I had done a couple of weeks before, or somebody called and I don't remember that they called.

But nothing major; and then, as time goes on, it gets better. Look at Vladimir Horowitz" — the concert pianist, who was also treated with ECT. "He was able to play billions of notes: Chopin, Tchaikovsky. The worst for me was that I'd forget something and my wife would say, 'Oh, I told you a couple of weeks ago but you didn't remember: you had the ECT.'"

What patients like Andre are complaining about is something more serious. They argue that ECT can result in wholesale amnesia, along with a steep decline in IQ. At the age of twenty, Andre says, her IQ was 156. Three years after her ECT it was around 112, and it does not appear to have increased since. Whether or not this is a result of ECT is hard to determine. Norman Endler, a psychologist who was himself treated with ECT, and Emmanuel Persad, a psychiatrist, wrote in their book *Electroconvulsive Therapy: The Myths and the Realities* (1988), "There is no conclusive proof that ECT causes permanent brain damage." What muddies the issue is that mental illness itself can cause cognitive defects, including a drop in IQ and in the ability to retain new memories. The informed-consent document for ECT used by Charles Kellner, a professor of psychiatry at the Medical University of South Carolina and the editor of the *Journal of ECT,* although scrupulous in its delineation of even the most severe side effects, states, "In part because psychiatric conditions themselves produce impairments in learning and memory, many patients actually report that their learning and memory functioning is improved after ECT."

In some cases a profound deterioration of cognitive ability is clearly the result of mental illness. Harold Sackeim, the chief of biological psychiatry at the New York State Psychiatric Institute, in New York City, and probably the most prolific ECT researcher in the world, told me about a colleague whose son had a psychotic break while a student at Harvard and now can't hold down a job at a fast-food restaurant.

Andre, Sackeim says, has shown him her medical records; he says that she may have experienced a similar breakdown. But there is no way to know for sure whether ECT was the culprit in Andre's loss of IQ and memory. "In very rare cases," Sackeim acknowledges, "there will be profound memory loss. People can lose years of their lives."

Jeremy Coplan, a professor of psychiatry at SUNY Downstate

Medical Center, in Brooklyn, who, like many other psychiatrists, doesn't actively treat with ECT but does refer patients for it, told me that the issue of memory loss is, unfortunately, often downplayed by psychiatrists. "For instance, someone may forget where the bathroom is in their house — at least temporarily," he said. "There can be a profound disruption of memory — not a minor thing if you put yourself in the patient's shoes." But, he said, it's a matter of risk versus benefit. "It's better that the patient is temporarily disoriented than seriously depressed for years."

The effect that ECT has on memory has been notoriously hard for ECT practitioners to concede. "The field has been under attack for such a long period of time," Sackeim says, "that a defensive posture was developed where limitations of the treatment were not acknowledged. So people complained of profound cognitive effects, and [those effects] were attributed to an ongoing psychopathology and essentially dismissed. I think that hurt the field of ECT."

Lately doctors have been taking special pains to spell out the risks that patients face. "I tell all my patients that they are going to have memory loss," Sackeim says. "In the vast, vast majority of patients that will be limited to a few months surrounding the course of treatment. There will not be a blank slate. But there will be gaps in memory. And the vast majority of patients say that's a small price to pay for getting well. It's not really a big deal to them. But I also tell them that in very rare instances it can be more extensive, and that no one can tell for certain who is going to experience that and who is not."

It has taken some time for a full disclosure to seep into the official literature. The report published this year by the APA Committee on ECT contains that organization's first substantial discussion of the possibility of serious memory problems.

There was a moment at the 1985 NIH conference, Peter Breggin recalls, when patients who had had positive experiences with ECT were asked to step up to the lectern and tell about their illness and recovery. Breggin, who is the director of the International Center for the Study of Psychiatry and Psychology, in Bethesda, Maryland, had already delivered a lecture titled "Neuro-pathology and Cognitive Dysfunction from ECT," and he listened intently as the patients spoke. Afterward one of them pressed a note into his hand, thanking him for speaking out about the side effects of ECT. "This was one of the *pro*-ECT people," Breggin told me when we spoke recently. "They were up there to tell people that ECT works, and here this person was thanking me for providing a dissenting opinion."

For Breggin, the experience epitomized ECT's ability to reduce patients to docility — to the point where they are willing to praise a treatment they feel has done them harm. In his view, ECT is a purposeful assault on the brain. He has been publicizing this opinion since 1979, when his first book, *Electroshock: Its Brain-Disabling Effects,* was published. Since then Breggin, a psychiatrist by training, has made a career out of attacking psychiatry and its methods. He has written several books arguing against the use of medication to treat mental illness, and he claims to be responsible for quashing the resurgence of lobotomy. His most recent efforts have been directed at establishing a link between antidepressants and the Columbine massacre. When Breggin discusses psychiatry, it is in the brusque manner of an aggressive debater.

Though Breggin has waged many campaigns, he has attacked ECT particularly vehemently, arguing that it causes "severe brain dysfunction" and that it creates in patients profound feelings of apathy or delirium. Psychiatrists welcome either outcome, he told me, because they can note with satisfaction on their charts that the patient is "complaining less" or has "an elevated mood." In this way, he says, psychiatrists fool themselves into believing that they are helping a patient when they are really doing harm. In his book *Toxic Psychiatry* (1991), Breggin wrote,

> If a woman received an accidental shock in her kitchen, perhaps from touching her forehead against a short-circuited refrigerator, and fell to the floor convulsing, she'd be rushed to the local ER and treated as an acute medical emergency. If she awoke the way a shock patient does — dazed, confused, disoriented, and suffering from a headache, stiff neck, and nausea — she'd be hospitalized for careful observation and probably put on anticonvulsants for months to prevent another convulsion. But on a psychiatric ward she'd be told she was doing fine and "not to worry," while the electrical closed-head injury was inflicted again and again.

Breggin first encountered ECT in the 1950s, when, as an undergraduate at Harvard, he volunteered at a state psychiatric hospital.

He was horrified, he recalls, at the conditions on the hospital's "back wards." Schizophrenic patients were left mumbling and rocking back and forth, without any human contact. They were led, zombie-like, to be treated with insulin coma or ECT. Breggin believed that if the patients were exposed to a more empathic environment, and one that provided for their basic needs, they would get better, so he persuaded the hospital administration to start a program of "love and care." He contends that plain old kindness worked. Later, as a resident in psychiatry and a teaching fellow at Harvard Medical School, Breggin observed firsthand the trend in psychiatry away from psychotherapy and toward physiological treatment, and he found it very disturbing.

"Mental illness," he says, "is a metaphor. It's not reality. When patients come into my office and say that they're depressed, I don't give them medication. I ask questions: What is their life like? What is their story? Where are they from? How did they get depressed? Why do they *call* it depression? Depression isn't caused by some mythical biochemical imbalance. It's another word for hopelessness."

This is a philosophy that Breggin absorbed from his training under Thomas Szasz, one of the forerunners of the "anti-psychiatry" movement. In the 1960s — along with Erving Goffman, R. D. Laing, and Michel Foucault — Szasz, a refugee from Nazi-era Hungary and a psychiatrist, promoted the view that mental illness is a social construct. Breggin's language is taken straight from his teacher. In the revised edition of his 1961 book *The Myth of Mental Illness,* Szasz wrote, "'Mental illness' is a metaphor. Minds can be 'sick' only in the sense that jokes are 'sick' or economies are 'sick.'"

Breggin is scorned by mainstream psychiatrists for his links to Szasz and for his contemptuous attitude toward physiological psychiatry. "Lots of fields have splinter groups," Harold Sackeim says. "Increasingly the dominant perspective in psychiatry is a biochemical one. There are people who, on ideological grounds, feel that this shouldn't be the case. They think psychotherapy should be the first line of treatment." But, he says, this opinion isn't necessarily benign. "Breggin will argue that a cup of tea, chicken soup, and a lot of hugging will get a psychotically depressed patient well. And he'll kill a lot of patients that way. That's why he doesn't have hospital privileges."

Still, Breggin has hit a nerve. Patients who have had negative experiences with ECT restate his arguments almost verbatim. By demonizing psychiatrists, by "exposing" their claims, Breggin has suggested answers to patients seeking to understand why they continue to suffer.

If practitioners of ECT tolerate "survivor" groups and disdain dissenting psychiatrists, they actively loathe the Citizens Commission on Human Rights. The inside of the pamphlet I have — one of many published and disseminated by CCHR — is an indication of why. A quick sampling of chapter headings: "Perpetuating Cruelty," "Therapy or Torture?," "The Nazi Heritage" ("electroshock's development . . . traces back to a dark alliance between psychiatry and the Nazi concentration camps"), "Apartheid and ECT," "ECT Promotes Breast Cancer," "Shock from Birth to Grave." Bolts of electricity in vivid neon colors provide visual unity here, emanating from the heads of pregnant women, fetuses, piglets. CCHR does not believe in subtlety.

The commission maintains offices in forty states and chapters in thirty other countries. It has used its branches in part to lobby for legislation against ECT. In 1974 it worked to get the California legislature to prohibit ECT for patients under the age of twelve. It has several times been instrumental in introducing legislation in Texas to ban ECT altogether. Although the legislation has failed, Texas is now, owing in large part to CCHR's efforts, the state in which it is the most difficult to get the treatment. Recently CCHR supported a bill in the Italian region of Piedmont which succeeded in banning ECT for children, the elderly, and, in most cases, pregnant women. That CCHR has effectively and perhaps permanently damaged the public image of ECT is one of the few things about which the commission and psychiatrists agree.

CCHR was founded in 1969 by the Church of Scientology, which by now has a fashionable Hollywood aura — John Travolta, Tom Cruise, and Nicole Kidman are all members. Scientology, "an applied religious philosophy," seeks to change the world through a system known as Dianetics, a term made familiar by a series of TV commercials for a book of the same name by the late L. Ron Hubbard, Scientology's founder and a science-fiction writer. Through Dianetics, Scientologists hope, according to the church's Web site, to create a utopia "without insanity, without criminals, and without war, where the able can prosper and honest beings can have rights, and where man is free to rise to greater heights." In CCHR's view, one of the greatest threats to this vision is abuses inherent in psychiatry, which damages the mind instead of soothing the soul. "For more than 115 years, psychiatrists have treated man as an animal," CCHR's Web site states. "They have assaulted, sexually abused, irreversibly damaged, drugged, or killed, all under the guise of 'mental healing.'"

CCHR was co-founded by Thomas Szasz, and its members take pains to emphasize this fact. Their connection to "the Church," as they call it, is spoken of less frequently. CCHR is separately incorporated, and although virtually every CCHR member worldwide also happens to be a member of the Church of Scientology, this is by choice, the organization says, not by compulsion. Rather than promote Scientology, CCHR seeks to lay out the evidence of psychiatry's misdeeds through the use of statistics, anecdotes, journal articles, news accounts, and hospital records.

The most voluminous resource for anti-ECT information within CCHR is Jerry Boswell, the director of the commission's Texas branch and the man most responsible for the state's stringent ECT laws. Boswell is patient and even-tempered, and his voice — soft and deep, with a heavy drawl — conjures the image of a large man in boots and a cowboy hat. At one point in a recent phone conversation with him I mentioned the TV personality Dick Cavett, who has very publicly and very positively spoken about how ECT helped him out of a terrible depression. "With ECT you have to ask the question of how much electricity was used," Boswell said. "Let's say you have Dick Cavett on your couch. Are you going to shock him at three hundred percent above the seizure threshold, or are you going to give him less electricity? You're going to give him less, because he's a public figure."

CCHR continually alleges that ECT uses "too high" a level of electricity. This has been difficult for psychiatrists to counter, because the very concept of "too high" leads immediately into contentious terrain. Dozens of studies have been done to determine how much electricity produces the most therapeutic seizures. On the basis of these studies some researchers have recommended that ECT devices be equipped to deliver *more* electricity. A 1991 paper by Harold Sackeim, "Are ECT Devices Underpowered?," published in the *Journal of ECT* (then called *Convulsive Therapy*), questioned the ability of contemporary devices to stimulate an ideally therapeutic seizure.

Whatever damage CCHR may have done to ECT, the organization has unquestionably improved the gathering of statistics regarding the treatment. The results, however, have not been advantageous to CCHR's cause. Several years ago CCHR lobbied successfully for compulsory reporting of ECT cases in Texas. William Reid, a clinical professor of psychiatry at the University of Texas Health Science Center, in San Antonio, and three other authors recently published in the *Journal of Clinical Psychiatry* all of the center's available data from September 1993 to April 1995. The article reported that 97.5 percent of all admissions were wholly voluntary; that the percentage of patients exhibiting "severe" symptoms was reduced from 70.7 prior to ECT to 2.4 afterward; that the percentage of patients with "moderate," "severe," or "extreme" memory dysfunction decreased after ECT; and that no bone fractures, heart attacks, or deaths occurred during treatment. Of the 2,583 patients described by the data, eight died within two weeks of their last treatment, but only two of these deaths may have been related in any way to ECT. The authors write,

> We are aware that anti-ECT groups have used the publicly available . . . data to support their contentions that ECT is dangerous and unnecessary and to campaign in the Texas legislature to ban the treatment altogether. We believe that those groups have often misinterpreted and/or misused the . . . data. We hope that this paper promotes objective discussion among clinicians, patients, families, and those who influence patients' access to this important treatment modality.

McLean Hospital has the sprawling lawns and architectural mien of a small New England college. Its forty-two buildings, almost all made of brick, are spread out over 242 acres. Adirondack chairs grace the lawns. Even early in the morning people are strolling about, and it is impossible to tell which are patients and which are staff members.

As at most hospitals, ECT at McLean is administered early on Monday, Wednesday, and Friday mornings — a cycle that allows patients to spend at least two days resting between treatments. In a typical year doctors at McLean give about two thousand ECT treatments to about two hundred patients. The diagnosis for almost all of them is some form of acute depression. Most have experienced what psychiatrists gently call "suicidal ideation." On the April morning that I visited to watch a treatment, Michael Henry, the head of McLean's ECT programs, was scheduled to treat sixteen patients, all of whom fit into those two categories. Henry seems to display all the qualities one hopes for in a psychiatrist. He has soft, comforting features; indomitable patience; and a voice that remains calm even when the situation calls for some emotion.

I arrived at the hospital before eight A.M. and was met in the reception area by a staff member in the hospital's public-affairs office. (This was the first time that a reporter was to be allowed to watch an ECT procedure at McLean.) A few minutes later I was shown into the treatment room, which looked like a small operating room but was less intimidating. With the middle of the room dominated by the table on which the patient lay, there was little space for the small crowd that had assembled: Henry, the anesthesiologist, a nurse, a third-year medical student, another staff member, and me.

The patient appeared to be in his late fifties, with gray hair and a touch of stubble. He was wearing jeans, a purple long-sleeved shirt, and white tennis shoes. He seemed unalarmed by the treatment that was to come, but his countenance betrayed the anguish of what Henry had told me was a depression whose manifestations included somatic hallucinations — illusions of movement and disease in different parts of the man's body. A year earlier the patient had gone through a course of ECT for similar episodes. That course had shown positive results, but the patient had recently relapsed and opted for more ECT. The treatment he was receiving that morning was his sixth in this course. I later asked Henry how many the man was to have. "That depends on him," he said. "We let the patient decide. We are very reluctant to push ECT."

The treatment began when the anesthesiologist injected a muscle relaxant and a general anesthetic into the patient's arm. The nurse inflated a blood-pressure cuff around his right ankle, which would prevent the relaxant from reaching his right foot and thus would provide a place where Henry could observe muscle contractions. She gently rubbed his hand as he went under. The anesthesiologist fit a plastic mask attached to a turnip-shaped bag over the patient's mouth and proceeded to squeeze oxygen into his lungs. Manually aided respiration has become standard procedure in ECT; it helps the patient not only to breathe once the muscle relaxants have paralyzed his diaphragm but also to rise from the anesthesia with a minimum of discomfort and memory loss.

Henry rubbed conductive jelly on two electrodes and placed both on the left side of the patient's head. Unilateral ECT, as this is called, is now the most common form. For years researchers debated whether this method was less effective than bilateral ECT, which involves placing one electrode on either side of the head, thus causing the seizure to affect both hemispheres of the brain. It has recently become clear that the difference in effectiveness is negligible but that unilateral ECT causes much less serious aftereffects.

Henry walked over to the ECT device, which looks like a large stereo receiver, and pressed a button. The patient's right foot seized, as though experiencing a sudden itch or a slight muscle spasm, and after ten seconds that was it. The procedure was gracefully mundane — anticlimactic, I couldn't help thinking. As we walked out, Henry said, "We try to keep it as absolutely boring as possible. The less interesting the better."

He could have been speaking for nearly all his fellow practitioners of ECT. Henry understands full well that the treatment's reputation is more complicated. Despite all the improvements in patient care, despite all the subtle tweaks and the impressive monitors affixed to the devices, ECT, Henry says, is still fundamentally the same treatment it was sixty years ago. The theory has remained fixed: shock a patient with enough electricity so that he'll have a seizure, and he'll probably get better. It's a blunt idea, medically speaking, and when pills that silently alter neurochemistry are the frame of reference, it is tough to warm up to something so primitively straightforward — even if for some reason it seems to work.

A number of ECT's most dedicated practitioners express a distaste for engaging in public efforts to bolster its reputation. One reason they give is that such undertakings would require pressing patients into service as witnesses. "We are here to do good by patients," Henry told me, "not to create poster children." In any event, among ECT practitioners there is considerable apprehensiveness about the media. In 1995 *USA Today* ran a three-part story about ECT that began with the death of a seventy-two-year-old woman during treatment; understandably, the article's publication had serious repercussions for patients' willingness to undergo ECT. In 1980 the *Atlantic Monthly* ran an article titled "Electroshock: The Unkindest Therapy of All," which Max Fink likened to *Mein Kampf.*

A skeptical press is symptomatic of a larger phenomenon. Psychiatrists assume that anti-ECT activists represent a fringe viewpoint on mental illness, whereas the evidence suggests that the anti-ECT outlook is actually close to the public's. In 1999 the Office of the Surgeon General released its first ever report on mental health. The report cited estimates that two-thirds of all cases of mental illness in this country go unreported. One of the main reasons the report gave for this is a widespread disbelief in the biological origin of psychiatric disorders. Despite the fact that major depression ranks second only to heart disease in the nation's "disease burden" (a measure that takes both mortality and morbidity into account), and despite the great scientific leaps that psychiatry has made, the report found the stigma associated with mental illness to be overwhelming: many people do not even accept that mental function is the work of a physical organ — a basic tenet of psychiatry. This suggests that the main obstacle ECT proponents face may be not proving its inherent usefulness but proving that the brain is an organ like any other, capable of breakdown.

When the surgeon general's report came out, it included a statement about ECT: "First-line treatment for most people with depression today consists of antidepressant medication, psychotherapy, or the combination . . . In situations where these options are not effective or too slow . . . electroconvulsive therapy (ECT) may be considered." This wasn't the original wording. Two months earlier a consumer rights activist had leaked the section dealing with ECT, which had called it a "safe and effective treatment for depression." A torrent of protests flooded the surgeon general's office. CCHR sent a sixteen-page document denouncing what it saw as a categorical endorsement of ECT. Linda Andre held meetings with an administrator working on the report. In the end the statement was softened.

However, the central message of the report — that there exists an enduring, peculiar, and unfortunate double standard involving the "physical" and the "mental" illnesses — was not mitigated. The predominant belief in the United States, the report indicated, is that it is all right to be subject to infection, degeneration, and microscopic revolts from the neck down. But a moral culpability is attached to whatever afflicts our minds. The double standard extends to treatment. We concede that coping with diseases of the body may of necessity bring about painful, even dangerous, side effects. We concede that we must weigh risks and benefits. But with psychiatric treatments, especially ECT, any possibility of harm is deemed wanton and intolerable. The discrepancy in attitudes is a strange one. According to Joseph Coyle, the chairman of the Department of Psychiatry at Harvard University, 15 percent of severely depressed patients commit suicide. It is a lethal disease. ECT doctors often draw a parallel with cancer: the treatments for cancer can be as damaging as the disease itself, they point out, yet there are no anti-chemotherapy lobbyists.

More important than questioning why anti-ECT lobbyists persist is asking what psychiatrists might do to counter the criticism. The answer from some is that they are already doing all they need to do. ECT use seems to be on the rise, even if slowly, and psychiatry's professional organizations are continually refining treatment guidelines. Greater advocacy efforts seem not to be on anyone's agenda, perhaps for fear of luring ECT's detractors into even louder denunciations.

There is still the possibility that a more benign method will be found to produce therapeutic seizures in the brain. Clinical trials are under way at hospitals worldwide for a treatment known as transcranial magnetic stimulation, which in one of its forms uses a strong magnetic field to create a seizure that is much more precise in intensity and placement than an ECT seizure. Convulsive TMS could drastically reduce memory loss, and thus could be an advance in convulsive therapy as marked as the move from Metrazol to electricity, sixty years ago. But it is likely to be years before TMS is fully developed and finds its way into treatment rooms around the country. In the meantime, patients must continue to seek out hospitals that offer ECT. And ECT will continue to offer benefits that other treatments do not.

As for Michael Henry's patient, he underwent six more treatments and was released, in good condition and well oriented.

PETER STARK

The Sting of the Assassin

FROM *Outside*

"SMELL THIS!" Mary said, pressing the flower to Gil's nose. "I wish we'd had some of these at the wedding."

He sniffed cursorily and turned his face away. "Let's go find a beach."

"Why don't we stay here awhile first?" Mary said. "We can find beaches anytime. But this" — she waved the flower up toward the leafy, cathedral-like rain-forest canopy, the thick vines dangling from the arching branches, the orchids sprouting from tree trunks, the tangled profusion of life — "this place is extraordinary! It looks like *The Land Before Time.*"

"I'm heading back to the car," Gil said.

"How can you expect to appreciate the beauty of a place if you refuse to spend some time in it?"

"I don't consider this place beautiful," Gil replied. "I think it's malevolent. And besides, there are ants. Let's go."

Mary dropped the flower. "You always want to be somewhere else, Gil," she said. "Why can't you just enjoy where you are?"

But he'd already started to walk back along the overgrown path, angrily shaking his sandaled feet to dislodge the crawling ants.

The honeymoon had been rough. They seemed to fight at least once every day, each blowup followed by a long, dripping silence. It hadn't been like this back in the States. They'd met at a bird-watching camp on the New England coast when her first marriage had just ended and his was about to end, the victim of too much time devoted to his law practice. By the time the trip was over, they'd arranged their first "date," a four-day snorkeling expedition to the

Bahamas. He proposed to her at sunset on the fourth day, and they married just three months later. It seemed only appropriate that for their honeymoon they should choose one of the world's most exotic places: the Cape York Peninsula, a lush spit of land that projects like a sharp spine from the northeast coast of Australia into the tropical waters of the Coral Sea.

They drove in silence along the coastal highway. To the right, through breaks in the low forest that fringed the beach, they caught glimpses of the sea — calm and blue-green. "When do you want to go out to the reef?" Gil asked finally. It was his way of offering to make up.

"How about tomorrow?" Mary replied. "We could go to the beach today and tonight ask the hotel desk clerk to arrange a boat."

It was her way of accepting.

Gil slowed the car where a sandy track led from the road toward the water. "How about here?"

"Oh, Gil, it's perfect!" The curving strip of yellow sand glistened and the water radiated a vivid aquamarine. They climbed out of the car and walked down to the beach. They couldn't see another sign of human life along the three-mile crescent. Gil wrapped his arms around Mary and hugged her.

"I'm so happy to be here with you," she whispered into the soft blue fabric of his T-shirt. She was quite sure she meant it.

"Me, too," he said into her sweet-smelling, honey-blond hair. They'd get used to each other eventually.

They dropped their packs, pulled out their towels, and spread them on the warm sand.

"The water looks lovely," she said. "Let's go for a swim."

"I don't know about swimming here," Gil replied, sitting down abruptly on his towel. "There aren't any enclosures."

Back at the popular beaches near their hotel they'd seen people swimming inside what the locals called "stinger nets." These were floating, corral-like enclosures made of fine netting designed to keep out jellyfish. As they strolled along the beach, they'd read the sign posted at one enclosure: WARNING: MARINE STINGERS ARE DANGEROUS OCTOBER TO MAY.

"I saw people swimming outside the enclosures," Mary said. "They say the worst of the jellyfish season is over."

"I still don't think it's a good idea," Gil said.

"Gil, how can you let this incredible water go to waste? Here we are in paradise and you're sounding like a lawyer."

"I'm just trying to be prudent in a place where we don't know the score."

She stripped off her shorts and T-shirt and let them drop to the sand. Underneath she wore an aqua bikini. She started walking toward the water, propelled as much by her defiance of his oppressive caution — of his whole being — as by the tempting blue sea.

"You're being foolish, Mary."

When it comes to predatory animals, humans have little to fear but themselves. We kill one another at a rate of more than one million per year, mostly wartime casualties. The second-deadliest threat is snakes, which kill over 100,000 people annually, followed by crocodiles (960), and tigers (740). The much feared shark falls far down the list — only about seven human victims annually worldwide — making it a lightweight compared to the ostrich, which when cornered can kick viciously with hammer-like feet and sharp talons and kills some fourteen people every year. As for the ferocious grizzly bear, it ranks about the same as mustelids (weasels, badgers, skunks, and the like), which kill an average of four humans a year, primarily pet ferrets attacking unattended babies. A couple more reassuring statistics: in the United States and Canada, you have more to fear from moose (six deaths per year) than any other creature except snakes (twelve). And the most likely place in the States to be attacked by an alligator (three deaths between 1992 and 1998) is not deep in some swamp, but on a golf course.

Very few animals stalk humans as prey, and those that do, such as the infamous man-eating tigers of India and lions of Africa, tend to be individual animals — typically outcast young males searching for new territory — that lose fear of humans and develop a taste for their flesh. Animals, for the most part, attack humans only when they are surprised or feel threatened or when defending their offspring. Snakes kill far more humans worldwide than any other animal, but as one authority states, they "have never been shown to attack without provocation, despite lengthy historical commentary to the contrary." Most bites occur in tropical countries when a rural villager unwittingly disturbs a snake, often by stepping on it in the dark. In the United States, by contrast, many victims are under the influence of alcohol or drugs and, in the words of another expert, are "messing with" the snakes. Researchers in Alabama have noted a statistical drop in venomous snakebites among adult men when University of Alabama or Auburn University football games are televised, presumably because the men are ensconced in their TV rooms.

There are no statistics showing that one region of the world is more dangerous than another in terms of animal attacks. Still, one can speculate. It would seem that parts of Africa inhabited by big game would make the list, as would the snake-infested Amazon Basin and parts of Southeast Asia such as Vietnam's Mekong Delta, where in one area studied by researchers, cobras, kraits, and vipers kill more than 2,700 people each year.

On the list, too, one expects, would be the coast of Australia and the region around Cape York Peninsula, which is home to an assortment of sharks, venomous snakes, poisonous fish, and the deadly saltwater crocodile. Of all those faunal hazards, one creature, despite its diminutive size, towers above the rest. This is a graceful-looking jellyfish not much larger than a grapefruit, known by the scientific name *Chironex fleckeri,* generically called a box jellyfish. As writer Bill Bryson noted in his recent best-selling book about Australia, *In a Sunburned Country, Chironex* is perhaps the most lethal venomous creature on earth. At least sixty-three *Chironex* deaths have been recorded along the coast of northern Australia since 1900, though many more deaths have probably gone undocumented through other parts of the South Seas. By some estimates, the venom of a *Chironex* can kill a human in less than a minute.

Mary waded deeper. The rippled sand massaged the bottoms of her feet and the warm tropical water soothed her skin like a mineral bath. The sea extended like a huge, placid lagoon toward the Great Barrier Reef, some thirty miles offshore. She badly wanted to see the reef, its spectacular corals and fish. It was hard enough to convince Gil simply to go for a swim; how difficult would it be, despite their plans, to actually drag him out there? She kept wading. She knew she wasn't being prudent, but the water looked fine, and his annoying, persistent caution made her want to go deeper, away from him. Was this marriage a mistake? she wondered. Already the fights she had with Gil resembled the same shopworn arguments she'd had with Tom, her ex-husband. If Gil turned out to be like

Tom, cautious and unimaginative, she'd leave him. Maybe first have the child she wanted, and then go.

"Come on in," Mary shouted, up to her hips in the clear water. "See, there's nothing to be scared of."

Gil waved her off. "I'm perfectly happy sitting here," he said.

But he wasn't happy. How could he be happy with someone who was pushing him all the time? A constant, tiny shove-shove-shove. This was supposed to be a vacation, wasn't it? He pushed himself hard enough at work; he didn't need someone needling and telling him to loosen up. Mary now reminded him of his ex-wife, Betsy, who for five years had chided him to work less, travel more, go for hikes and picnics and visits to galleries. She was relentless. He'd finally moved out. Then, strangely, he found himself taking up some of the activities he'd resisted for so long, as if to prove to her that he wasn't the stodgy person she thought. It had been a great relief to meet Mary at the bird-watching camp — she was so patient and understanding when he told her about his failed marriage — but now it seemed he was reliving those five bad years. Couldn't she just relax a bit instead of charging from rain forest to reef to outback? He didn't want to think about how it would be when they got back home. He'd already made the mistake. How long would a divorce take? And, more important, what would it cost him?

"Last chance!" she called, twisting her head toward him.

"Hurry up and swim if you're going to swim," he answered, his irritation rising. "Otherwise let's go back to the hotel."

She brought her arms over her head and sprang gently off the sand with her toes. As she dived underwater, sliding in with a gentle splash, she made her decision: *That's it. We're finished.*

About a year earlier, late in the October–April wet season, two spawning *Chironex fleckeri* released sperm and eggs into a river estuary not far from Gil and Mary's isolated beach. Joining in the warm water, the two cells soon grew into a minute ball — the planula, as this stage is called — and dropped to the river bottom, attaching to the underside of a rock. There the planula sprouted the beginnings of a crown of tentacles and grew into a tiny polyp that by the end of the dry season had metamorphosed into a small jellyfish. Just ahead of the monsoon rains, the *Chironex* propelled itself out of the estuary and into the Coral Sea.

For the next few months it gently pulsed through calm waters along the coast, feeding and growing, avoiding violent currents and waves that could tear its delicate tissues. Its body, an almost transparent milky white, 95 percent water, developed into a graceful bell shape with a squarish, four-cornered bottom rim — thus the name box jellyfish. By filling its bell with water and squeezing it out like an umbrella opening and closing, the jelly could jet along at speeds up to four miles per hour. A limb shaped like a chicken's foot dangled from each of the four corners of the bell's rim and from these grew the jelly's tentacles, as many as fifteen per limb. Only a quarter of an inch in diameter and stretching more than fifteen feet when fully extended, the tentacles resembled twisted lengths of skinny — and highly charged — electrical conduit.

Using primitive eyes to help it spot large objects, the jelly followed schools of shrimp that congregated just off the sandy beaches. Sensitive to strong sunlight, it lingered near the sea floor during the height of the day, rising toward the surface as the sunlight weakened in late afternoon, trailing its long tentacles behind it, trolling for prey. When a shrimp or a small fish came past, inadvertently brushing a tentacle, it died almost instantly. The *Chironex* would then reel in the catch, feeding the meaty morsel into wide, grasping lips.

Mary never saw it.

Plunging beneath the smooth surface, she glided underwater, savoring the trickling sound in her ears and the warmth of the tropical sea, relieved that she'd made the decision to leave Gil. Then something brushed against her arms. She flinched, startled. Was it seaweed? Then it brushed against her shoulders, her midriff, her back.

Reflexively, the jelly's tentacles contracted, piling themselves in loops onto her skin in an instinctive attempt to apply maximum surface area — and thus maximum venom — to the victim. Each tentacle was studded with millions of tiny venom sacs, called nematocysts, the entire jellyfish armed with an estimated five billion of them. The touch of Mary's skin — indistinguishable, as far as the jelly was concerned, from a shrimp or a fish — triggered the release of coiled tubules with sharp tips that sprang from the top of each nematocyst like a jack-in-the-box. The tubules, each about .03 inches long, jabbed their points into Mary's skin, injecting venom from the sacs.

Mary gasped. A stream of bubbles escaped from her mouth. Searing pain burned over her arms and torso. She kicked toward the surface, tearing at the stinging things adhering to her skin.

Gil, leaning back on his elbows and staring impassively through his tortoiseshell sunglasses, watched her graceful dive. The surface erupted with foam. He sat up. It was Mary, her arms flailing wildly. Her screams carried across the calm water.

He knew immediately what had happened.

"I'm coming!" he shouted, launching himself off his towel. He would be the heroic husband, vindicating himself for the caution that she had so derided. He flung off his $375 sunglasses and charged into the water. Then he suddenly stopped.

Would it sting him, too?

She was about fifty feet away, in water that was probably just over her head, struggling to put her feet on the bottom. Where was the jellyfish? It could be anywhere. How long were its tentacles? He had no idea.

"Where is it?" he shouted.

It was as if Gil, standing paralyzed in knee-deep water near shore, had also been injected with the jellyfish's powerful, stunning venom. He looked up and down the beach for someone who might help, but they'd chosen this beach precisely because no one was here.

"Can you swim to me?" he shouted.

She raised an arm helplessly, and he could see the whiplash-like welts already spiraling along it as she cried and gagged in the water.

No one knows exactly what's in the venom of *Chironex fleckeri* that makes it so potent. Research has been stymied in part because the venom is susceptible to heat, making it difficult to analyze without chemically altering it. Scientists believe that the venom, made up of protein-like substances, contains three major toxins: one that damages skin, one that affects the blood, and another that works, sometimes fatally, on the heart and other organs. The incredible skin pain, some have speculated, could be due in part to a compound called 5-hydroxytryptamine, found in the tentacles of many types of jellyfish as well as in bees and stinging nettles.

It has been estimated that *Chironex* venom enters the circulatory system of a healthy person within twenty seconds of the sting, as the tiny tubules inject thousands of doses of venom directly into the capillaries just beneath the skin. This is unlike snakebite, in which the snake's fangs leave a few large deposits of venom that take hours to be absorbed into the tissue.

Mary's "fight-or-flight" response — which might be useful if she were fending off a shark attack — had sent surges of adrenaline through her body, boosting her heart rate from 80 beats per minute to its maximum exertional rate of 180. Her rapidly contracting muscle tissues, which demanded oxygen, triggered a rush of blood from her heart and lungs to her limbs and back again. The needs of her muscles, induced by her panic, helped carry the venom from the capillaries beneath her skin straight to her heart.

The venom quickly did its work. Something in the venom — a cardiotoxin, as it's known — began to wreak havoc on her heart's electrical system, causing a big upsurge in the number of positively charged calcium ions entering the cardiac muscle cells. An excess of calcium ions inside these cells causes spasms in the carefully timed contractions of the heart, a bit like throwing a jug of water onto the circuitry of an electric motor.

Mary's heart abruptly lost its rhythm. The ventricles, the heart's two high-pressure pumping chambers, normally contract smoothly from the bottom up, squeezing blood out the top. But the *Chironex* venom triggered chaotic contractions originating somewhere in the middle of the ventricular walls and firing at an arrhythmic pace of 240 beats per minute. Instead of a powerful, coordinated stroke pushing out three ounces of blood with every beat, her heart dribbled less than a twentieth of that amount. Her blood pressure plummeted. The blood flow to her brain slowed to a trickle. The bright circle of sunlight and seawater and pain faded to a twilight pool.

Her head dropped in the water, bobbed up, dropped again. She wasn't screaming now, though her lesioned arms still waved vaguely, keeping her afloat.

"Mary! Mary!" Gil shouted.

She gave no indication that she heard him.

Gil had never seen anyone die before, but it was perfectly clear to him that she was almost dead. He knew he'd replay this moment for the rest of his life: Mary floating face down, arms stirring dully, hair gently fanning out on the water, while he stood frozen and simply . . . watched.

He moved one foot forward across the sandy bottom. Then the other. Suddenly he was churning through the water toward her, no longer caring what happened to him, as long as he acted. Thigh-deep . . . waist-deep . . . chest-deep. She lay floating only ten feet away. He shuffled two steps closer. He stretched out his right arm, reaching up over the surface of the water, above any stray tentacles. He touched her left hand, feeling the sticky softness of the strands that were wrapped around it. He pulled away. Oddly, he felt no stinging. Were they somehow spent?

They were not. He felt no sting because the tiny venom-filled tubes launched by the nematocysts didn't have enough power to penetrate the thick skin of his palm. The hair on the back of his hand also acted as a barrier to their penetration. Women and children, who have less hair, smaller bodies, and tenderer skin than men, are therefore more susceptible to *Chironex* stings. Even a thin layer of nylon can repel stings, as discovered by Australian surfers and lifeguards, who pull pantyhose over their legs and torsos when swimming in jellyfish-infested waters.

Gil reached out again and gripped her left wrist, between tentacles. Adrenaline pumping, he dragged her 122 pounds through the shallow water and pulled her up onto the beach.

He dropped to his knees beside her. Strands of tentacle clung to her torso, etching purplish brown weals on her skin. If she survived, the tissue could die, and she might be scarred for life with the whorled markings, as if tattooed by a heap of rope.

Gil remembered the sign posted at the beach near town: FLOOD STING WITH VINEGAR. Dousing tentacles with vinegar prevents the nematocysts from firing. Gil had noticed the jugs placed at regular intervals along that beach, but there was no vinegar here. He grabbed his towel and wrapped it around his right hand, plucking the tentacle fragments from Mary's skin.

The severity of a *Chironex* sting depends largely on how much tentacle has made contact with the victim; about six feet is considered the minimum for a lethal dose in an adult human. By some estimates, a full-grown *Chironex* contains enough venom in all of its tentacles to kill up to sixty people.

About fifteen feet of tentacle had made contact with Mary's skin. Gil rolled her over. He watched her sand-covered abdomen for the rise and fall of breathing. Nothing.

Years earlier Gil had taken a course in CPR and then later, a refresher course before their trip to the Bahamas. He pinched Mary's nostrils, placed his open mouth over hers, and exhaled once. Her chest rose slightly as the air went in, then fell. But what about her heart? He felt for a pulse in her neck, holding his trembling fingers as still as he could. Nothing. Her heart was still in wild spasms, pumping at over 200 beats per minute, but her blood pressure was too low to register a pulse.

Gil had to become both her heart and her lungs, at least long enough for her body to recover from the venom's shock and regain some of its equilibrium. This was not a simple task for one person to perform. He'd have to work quickly. Laying his hands on top of her breastbone, he compressed her sternum, then released and shoved again. After the first fifteen thrusts he gave her two more quick breaths, followed by another set of fifteen compressions. His rapid, steady thrusts kept blood moving up her carotid arteries and into her brain at one-third its normal flow — not a lot, but enough to deliver the crucial supply of oxygen to the brain's starving tissues.

Thirty seconds passed. A minute. The pumping action he forced on her heart began to wash away the poisoned blood, dispersing it throughout her body.

Another forty seconds passed. Sweat flew from Gil's face and arms. He didn't know how long he could keep this up.

Embedded in the roof of the right atrium is the heart's own pacemaker, a bundle of muscle fibers called the sinoatrial node that generates its own electrical charges. As it fired, it tried to send impulses through Mary's heart like a wave, ordering contractions of the exact proper muscle tissue in the exact proper sequence. But tiny holes in her cell membranes remained opened from the venom's effect, and the calcium ions passed too easily into the cells, keeping the muscle in spasms. For her heart to resume its normal rhythm through CPR, Gil would have to keep up the chest compressions long enough for the venom to degrade, thirty to forty minutes.

The second minute passed. Gil stopped again to probe for a pulse. He felt something. But what? He placed his hands on her sternum and resumed the frenzied pace of chest compressions punctuated by breathing. Another minute. He was panting so hard now he had to pause to catch his breath. It seemed that he'd been working over her for a very long time. Was this futile? He felt again for her pulse, trying to understand what he sensed in his fingertips. Then he heard a small gasp. He looked at her chest. Another gasp. He saw her rib cage fall slightly.

He thrust his ear to her chest. He could hear something. Was it the *lub-dub* of his own thumping heart or the sound of her heart valves opening and closing as they should?

"Keep going, Mary!" he shouted.

He lifted his head from her chest. Now her breath came in small gasps.

"That's right, breathe, Mary, *breathe!*"

Her limbs stirred, brushing against the sand as if she remembered, deep in her subconscious, that she was supposed to swim. Her head lolled back and forth. Her eyes were still closed. Gil knew he'd have to pick her up in his arms and haul her to the car, then jounce over the rutted beach road to the highway to town. They'd arrive in twenty minutes, and the hospital would surely have a ready supply of the box jellyfish antivenin. Quickly administered, the antivenin (made from the blood of a sheep that's been immunized to the venom) does much to ameliorate the jellyfish's sting. She could relapse — *Chironex fleckeri* victims sometimes show a brief improvement and rising blood pressure before suddenly expiring — though Gil didn't know this. They'd put her on a respirator if she needed it, and give her medicine for the pain that would return once she woke up.

What would happen then? he wondered. What would they say to each other? Would she in some way be a different person, not pushing him always? And would he be different, not hanging back? Could a brush against this tentacled creature that had no brain, no sense of good and evil, change a life, two lives, a marriage?

He jumped up. He ran to the pile of clothes they'd shed on the beach and grabbed his daypack with the car keys in it. He knelt beside her again.

"Come on, Mary, come on!"

He allowed himself to wish now, to hope. Gil knew that he'd done everything for her that was humanly possible, far more than he would have ever guessed he could do. As her chest continued to rise and fall, tears began to fill his eyes.

CLIVE THOMPSON

The Know-It-All Machine

FROM *Lingua Franca*

FLORENCE ROSSIGNOL has just finished using an on-line travel site to plan a package tour across Europe. The site has prompted her for a few facts about herself: her date of birth, her education and nationality, her occupation. She has typed in that she was born in 1945 and trained as a nurse. She has also volunteered the fact that she is claustrophobic. As far as on-line shopping goes, it looks like an everyday event.

Except this Web site is smart — unusually smart. It has been outfitted with a copy of Cyc (pronounced *sike*), artificial-intelligence software touted for its ability to process information with human-like common sense.

At one point, Cyc detects a problem: the proposed tour involves taking the Channel Tunnel from London to France; Rossignol is claustrophobic. The Web site notes that Rossignol "may dislike" the Channel Tunnel, and Cyc justifies the assertion with a series of ten related statements, including:

- Thirty-one miles is greater than fifty feet.
- The Channel Tunnel is thirty-one miles long.
- Florence Rossignol suffers from claustrophobia.
- Any path longer than fifty feet should be considered "long" in a travel context.
- If a long tunnel is a route used by a tour, a claustrophobic person taking the tour might dislike the tunnel.

At the same time, Cyc scours the list of various cities on the tour and takes special notice of Geneva, where one can visit the Red

Cross Museum. This time, Cyc's thinking features the following steps:

- The Red Cross Museum is found in Geneva.
- Florence Rossignol is a nurse.
- Nursing is what nurses do.
- The Red Cross Museum (organization) has nursing as its "focus."
- If an organization has a particular type of activity as its "focus" and a person holds a position in which they perform that activity, that person will feel significantly about that organization.

Bingo. The travel site tells Rossignol to make sure she catches the Red Cross Museum in Geneva — but for God's sake, don't take the Channel Tunnel.

Doug Lenat is pleased. Though this impressive display happens to be a promotional demo (Rossignol is a fictitious character), it is, he explains, genuinely representative of his invention's unique abilities. Most computer programs are utterly useless when it comes to everyday reasoning because they don't have very much common sense. They don't know that claustrophobics are terrified of enclosed spaces. They don't know that fifty feet can sometimes be considered a "long" distance. They don't even know something as tautological as "Nursing is what nurses do."

Cyc, however, does know such things — because Lenat has been teaching it about the world one fact at a time for seventeen long years. "We had to kick-start a computer, give it all the things we take for granted," he says. Ever since 1984, the former Stanford professor has been sitting in Cyc's Austin, Texas, headquarters and writing down the platitudes of our "consensus reality" — all the basic facts that we humans know about the world around us: "Water is wet"; "Everyone has a mother"; "When you let go of things they usually fall." Cyc currently has a database of about 1.5 million of these key assertions. Taken together, they are helping Lenat create what he calls the first true artificial intelligence (AI) — a computer that will be able to speak in English, reason about the world, and, most unnerving, learn on its own. Cyc is easily the biggest and most ambitious AI project on the planet, and by the time it's completed, it will probably have consumed Lenat's entire career.

Encoding common sense is so formidable a task that no other AI theorists have ever dared to try anything like it. Most have assumed it isn't even possible. Indeed, with Cyc, Lenat has tweaked the noses of legions of AI researchers who have largely given up on the rather sci-fi–like dream of creating human-like intelligence and have focused instead on much smaller projects — so-called expert systems that perform very limited intelligent tasks, such as controlling a bank machine or an elevator. "Doug's really one of the only people still trying to slay the AI dragon," says Bill Andersen, a Ph.D. candidate specializing in ontologies (information hierarchies) at the University of Maryland and a former Department of Defense researcher who has used Cyc in several Defense Department experiments.

For all his progress, Lenat still receives mixed responses from much of the academic AI community. Not only does Cyc's highly pragmatic approach fly in the face of much scholarly AI theory, but its successes have taken place at Lenat's Cycorp, which is developing Cyc as a for-profit venture. Incensing his critics, Lenat has published almost no academic papers on Cyc in recent years, raising suspicions that it may have many undisclosed flaws. "We don't really know what's going on inside it, because he doesn't show anyone," complains Doug Skuce, an AI researcher at the University of Ottawa.

Lenat, meanwhile, revels in his bad-boy image. He accuses academic AI experts of being theory-obsessed and unwilling to do the hard work necessary to tackle common sense. "They want it to be easy. There are people who'd rather talk about doing it than actually do it," he says, laughing.

Still, some skeptics think Lenat could benefit from more deliberation and less action. "It's kind of crazy," says the Yale computer science professor Drew McDermott about Lenat's ambition. "Philosophers have been debating common sense for years, and they don't even know how it works. Lenat thinks he's going to get common sense going in a computer?" Push Singh, a graduate student at the Massachusetts Institute of Technology who is building a rival upstart to Cyc, has his own doubts: "Lenat and his team have been going for fifteen years and have only one million rules? They'll never get enough knowledge in there at that rate."

Lenat plans to take a huge step toward silencing his critics this fall, when he begins to "open source" Cyc. That is, he intends for the first time to allow people outside of Cycorp to experiment with their own copies of Cyc — and train it in their own commonsense knowledge. Eventually he proposes to let everyone in the world talk to Cyc and teach it new information, to elevate its knowledge to a level of near omniscience. "It'll get to the point where there's no one left for it to talk to," says Lenat, his eyes twinkling mischievously.

But when everyone can speak to Cyc, what exactly will they tell it? Can Cyc — or any AI system, for that matter — truly embody the common knowledge of humanity, with all its many layers, its contradictions and ambiguities?

Since this is 2001, Lenat has spent the year fielding jokes about HAL 9000, the fiendishly intelligent computer in Arthur C. Clarke's *2001: A Space Odyssey.* On one occasion, when television reporters came to film Cyc, they expected to see a tall, looming structure. But because Cyc doesn't look like much — it's just a database of facts and a collection of supporting software that can fit on a laptop — they were more interested in the company's air conditioner. "It's big and has all these blinking lights," Lenat says with a laugh. "Afterward, we even put a sign on it saying, CYC 2001, BETTER THAN HAL 9000."

But for all Lenat's joking, HAL is essentially his starting point for describing the challenges facing the creation of commonsense AI. He points to the moment in the film *2001* when HAL is turned on — and its first statement is "Good morning, Dr. Chandra, this is HAL. I'm ready for my first lesson."

The problem, Lenat explains, is that for a computer to formulate sentences, it can't be *starting to learn.* It needs to already possess a huge corpus of basic, everyday knowledge. It needs to know what a morning is; that a morning might be good or bad; that doctors are typically greeted by title and surname; even that we greet anyone at all. "There is just tons of implied knowledge in those two sentences," he says.

This is the obstacle to knowledge acquisition: intelligence isn't just about how well you can reason; it's also related to what you already know. In fact, the two are interdependent. "The more you know, the more and faster you can learn," Lenat argued in his 1989 book, *Building Large Knowledge-Based Systems,* a sort of midterm report on Cyc. Yet the dismal inverse is also true: "If you don't know very much to begin with, then you can't learn much right away, and what you do learn you probably won't learn quickly."

This fundamental constraint has been one of the most frustrating hindrances in the history of AI. In the 1950s and 1960s, AI experts doing work on neural networks hoped to build self-organizing programs that would start almost from scratch and eventually grow to learn generalized knowledge. But by the 1970s, most researchers had concluded that learning was a hopelessly difficult problem, and were beginning to give up on the dream of a truly human, HAL-like program. "A lot of people got very discouraged," admits John McCarthy, a pioneer in early AI. "Many of them just gave up."

Undeterred, Lenat spent eight years of Ph.D. work — and his first few years as a professor at Stanford in the late 1970s and early 1980s — trying to craft programs that would autonomously "discover" new mathematical concepts, among other things. Meanwhile, most of his colleagues turned their attention to creating limited, task-specific systems that were programmed to "know" everything that was relevant to, say, monitoring and regulating elevator movement. But even the best of these expert systems are prone to what AI theorists call "brittleness" — they fail if they encounter unexpected information. In one famous example, an expert system for handling car loans issued a loan to an eighteen-year-old who claimed that he'd had twenty years of job experience. The software hadn't been specifically programmed to check for this type of discrepancy and didn't have the common sense to notice it on its own. "People kept banging their heads against this same brick wall of not having this common sense," Lenat says.

By 1983, however, Lenat had become convinced that commonsense AI was possible — but only if someone were willing to bite the bullet and codify all common knowledge by brute force: sitting down and writing it out, fact by fact by fact. After conferring with MIT's AI maven Marvin Minsky and Apple Computer's high-tech thinker Alan Kay, Lenat estimated the project would take tens of millions of dollars and twenty years to complete.

"All my life, basically," he admits. He'd be middle-aged by the time he could even figure out if he was going to fail. He estimated he had only between a 10 and 20 percent chance of success. "It was just barely doable," he says.

But that slim chance was enough to capture the imagination of Admiral Bobby Inman, a former director of the National Security Agency and head of the Microelectronics and Computer Technology Corporation (MCC), an early high-tech consortium. (Inman became a national figure in 1994 when he withdrew as Bill Clinton's appointee for secretary of defense, alleging a media conspiracy against him.) Inman invited Lenat to work at MCC and develop commonsense AI for the private sector. For Lenat, who had just divorced and whose tenure decision at Stanford had been postponed for a year, the offer was very appealing. He moved immediately to MCC in Austin, Texas, and Cyc was born.

Lenat began building Cyc by setting himself a seemingly modest challenge. He picked a pair of test sentences that Cyc would eventually have to understand: "Napoleon died in 1821. Wellington was greatly saddened." To comprehend them, Cyc would need to grasp such basic concepts as death, time, warfare, and France, as well as the sometimes counterintuitive aspects of human emotion, such as why Wellington would be saddened by his enemy's demise. Lenat and a few collaborators began writing these concepts down and constructing a huge branching-tree chart to connect them. They produced a gigantic list of axiomatic statements — fundamental assumptions — that described each concept in Cyc's database: its properties, how it interacted with other things. "We took enormous pieces of white paper," Lenat remembers, "and filled walls, maybe a hundred and fifty feet long by about eight feet high, with little notes and circles and arrows and whatnot."

Over the next few years, those axioms ballooned in number — eventually including statements as oddly basic as:

- You should carry a glass of water open end up.
- The U.S.A. is a big country.
- When people die, they stay dead.

The axioms aren't written in everyday English, which is too ambiguous and nuanced a language for a computer to understand. Instead, Cyc's "Ontological Engineers" — Lenat's staff of philosophers and programmers, who call themselves Cyclists — express each axiom in CycL, a formal language that Lenat's team devised.

Based on the sort of symbolic notation that logicians and philosophers use to formalize claims about the world, CycL looks like this:

```
(forAll ?X
  (implies
    (owns Fred ?X)
    (objectFoundInLocation ?X FredsHouse)))
```

This expression states that if Fred owns any object, ?X, then that object is in Fred's house. In other words, as Cyclists put it, "all Fred's stuff is in his house." (Of course, as with all Cyc's knowledge, this claim becomes useful only in conjunction with other truths that Cyc knows — such as the fact that a person's car or beachfront property is too large to fit in his house.)

Cyc's inventory of the world, however, is only one part of its setup. The other part is its "inference engine," which allows Cyc to deploy its immense store of factual knowledge. This engine includes Cyc's "heuristic layer" — a collection of more than five hundred small modules of software code that perform logical inferences and deductions, as well as other feats of data manipulation. One module, for example, implements traditional *modus ponens* logic: if Cyc knows a fact of the form "If X, then Y," and Cyc knows "X," then it will conclude "Y." Other modules have the ability to sort facts by, say, chronological order.

On the one hand, the inference engine is what actually gives Cyc its innate smarts; without it, Cyc wouldn't be able to do anything with the information at its disposal. But on the other hand, as Lenat emphasizes, a computer can have state-of-the-art powers of data manipulation and still be worthless from a practical point of view; no machine can help you reason about the real world if it does not have commonsense knowledge to work with. Data manipulation, in Lenat's view, is the comparatively easy part. It's the data themselves that are devilishly difficult.

From the perspective of computing power, commonsense knowledge presents an additional difficulty in its sheer mass. As Cyc's knowledge base grew, the program had to sort through thousands of facts whenever it tried to reason. It began to slow down. "If you're trying to talk about the weather today, you don't want Cyc worrying about whether spiders have eight legs," Lenat explains. So the Cyclists have created "contexts" — clumpings of like-minded facts that help speed up inferencing. If Cyc is told a fact about tour trips to Greece, for example, it begins with its existing knowledge about Europe, travel, trains, and the like. In this sense, Cyc's processing strategies are akin to human cognition; we can discuss any given topic only by ignoring "99.999 percent of our knowledge," as Lenat has written.

By the early 1990s, Cyc had acquired hundreds of thousands of facts about the world and could already produce some startlingly powerful results. For example, Cyc could search through databases to find inconsistent information, such as twins who were listed with different birth dates. Cyc didn't need to be specially programmed to look for that sort of error — it just "knew" the commonsense idea that twins are always the same age. Pleased with Cyc's progress, Lenat spun his venture off from MCC to form Cycorp, a freestanding company.

Still, teaching Cyc new knowledge remains an excruciatingly slow process, filled with trial and error. If the Cyclists make a mistake, or forget to articulate explicitly some aspect of a concept, Cyc can reach some wholly implausible conclusions about the world.

A few months ago, for instance, Charles Klein, Cycorp's thirty-three-year-old director of ontological engineering, was asking Cyc a few test questions when he discovered something odd. Cyc apparently believed that if a bronze statue were melted into slag, it would remain a statue. What had gone wrong? Why was Cyc making such a basic mistake?

After a bit of forensic work, Klein found the problem. The Cyclists hadn't completely distinguished the concepts of bronze and statue. Cyc had been told that bronze was a material that retained its essential property — its "bronzeness," as it were — no matter what state it was in, solid or liquid. But now Cyc was trying to apply that fact to the statue aspect of "bronze statue." Cyc hadn't been told anything about statues that would invalidate its conclusion; nobody had ever thought it necessary to tell Cyc, for example, that statues are only statues if they're more or less in their original form. It's common sense, sure — but who would bother to meditate on it? "Trying to think of *everything,*" Klein quips, "is quite daunting."

This is the chief problem that all Cyclists face: commonsense knowledge is invisible. It's defined as much by what we *don't* say as by what we *do* say. Common knowledge is what we assume everyone has, because it's, well, obvious. This is precisely what makes commonsense knowledge so powerful as an intellectual tool — but it's also what makes it so hard to identify and codify.

"There's no book that you could read about common sense," Lenat points out, "that says things like, If you leave a car unlocked for ten seconds, then turn around, the odds of it still being there are really high. But if you leave a car unlocked for ten years, the odds are really low that it'll still be there when you come back."

The example of the unlocked car seems to raise a troublesome theoretical question: is this common sense because it is a basic *fact* about the world that most everyone knows? Or is it common sense because it demonstrates sound *reasoning*? Just as important: is this sort of conceptual concern a genuine threat to Cyc, or just more ivory-tower quibbling?

In academe, Cyc has always been a black sheep. Everyone in AI knows about it, and virtually everyone views it with skepticism. Lenat's critics have lambasted him for lacking a coherent theory of how intelligence and knowledge representation work — and for rushing ahead with an ill-thought-out project. When they look at Cyc, they see nothing but an ad hoc jumble of facts about the world picked in an overly idiosyncratic way by Lenat's team. For them, making human-level AI requires a better theoretical understanding of human intelligence and the fundamentals of reasoning and representation — and those are areas that are still, many argue, in their infancy.

In a 1993 issue of *Artificial Intelligence,* several reviewers sharply critiqued Lenat's 1989 book about Cyc. Yale's Drew McDermott led the charge, arguing that it was impossible to build a commonsense database without solving such philosophical problems as "the nature of causality." "We've been thinking about things like that for millennia," he points out.

McDermott suspects that it may not yet be possible to represent real-world common knowledge in logical, orderly languages such as Cyc's CycL — or any other language, for that matter. After all, humans don't always store and manipulate knowledge logically or in language. "If you go through a room and you don't bump into things, is that common sense?" he wonders. Nils Nilsson, an AI pioneer and Stanford professor emeritus, shares that concern. "You can describe in words how to swing a golf club," he concedes. "But can that really tell you how to do it? We still don't really know how to represent knowledge."

Critics and fans of Cyc both recognize that the goal of producing a complete inventory of commonsense facts is almost embarrassingly open to theoretical objections. Because so many philosophical issues about how to represent knowledge remain unsolved, building large knowledge bases is "something of a backwater," according to Ernest Davis, a professor of computer science at New York University and the author of *Representations of Common Sense Knowledge*. Starting in the 1980s, for example, much of the excitement in AI began to center around the use of narrowly focused self-learning systems — like neural nets — to crunch enormous bodies of data. Such systems are intended to "learn" to recognize patterns on their own, instead of being painstakingly taught them by humans. "People like it because it's a lot faster," Davis explains. "You'll never be able to get commonsense AI out of it, but you can do some pretty neat things," such as develop programs that can recognize visual images.

Lenat, however, is unmoved. He bashes right back at the naysayers every chance he gets — often in searingly witty prose. In his response to the 1993 reviews by McDermott and others, he argued that theory-heavy AI experts were suffering from the Hamlet syndrome — unwilling to take action and stuck in a circle of philosophizing. Too many AI theorists, he sneered, were "researchers with 'physics envy,' insisting that there must be some elegant 'free lunch' approach to achieving machine intelligence."

For Lenat, having a watertight theory isn't necessary for building useful AI. Quite the contrary: he argues that building a body of commonsense knowledge can only be done in a down-and-dirty engineering style. You put together a bunch of facts, test the system, see where it breaks, fix the bugs, and keep on adding more knowledge. Each day, when Cyclists talk to Cyc, they discover new erroneous assumptions that Cyc has — or new information that it doesn't have. "It's iterative. You have to do it every day, keep at it," says the Cyclist Charles Pearce.

Lenat compares building up Cyc to building a bridge. "You know, you have a stream and you want to build a bridge over it —

you can either do your best, experiment, and build the bridge. Or you can work out the physics of bridge building," he says. "It's only in the last hundred years or so that the theory of bridge building has been understood. Before that, it was almost like apprenticing to a master. There would be someone who would just intuitively know how to build a bridge — and every now and then, he would build one that would fall down."

Lenat's critics also complain that by developing Cyc in the private sector, he is forced to keep his cards too close to his chest. Aside from publishing his now dated 1989 book, he and his staff have produced barely a handful of papers about Cyc in academic or trade magazines. In fact, many AI academics — even Lenat's supporters — say they don't know enough of what's going on inside Cyc to be able to approve or disapprove. "He doesn't publish, because he doesn't need to," says one critic. "When I suggest things they ought to do, they just say, Well, we're not being paid to do that."

Compounding this resentment is the fact that in the world of large-scale knowledge bases, Cyc is the biggest — perhaps the only — game in town. As a result, some academics worry that Cyc's example discourages other AI researchers from building competing knowledge bases. Lenat's seventeen years of painstaking labor have shown exactly how difficult and involved the project is. For his part, Lenat quietly relishes the monopoly he has on commonsense knowledge. Some of his funders, including the Department of Defense and GlaxoSmithKline, already use experimental versions of Cyc to help "scrub" data in databases, cleaning out data-entry errors. Ultimately Lenat hopes to license Cyc to all software makers worldwide — as a layer of intelligence that will make their systems less brittle. It could become like "Intel inside," he suggests, or the Microsoft Windows of the AI world.

In the meantime, Lenat cheerily admits to having "almost no relationship" with the academic world. It has rejected him, he argues — not the other way around. Ten years ago, he sent some of his staff to speak, on invitation, at a major academic conference on common sense. Their papers were "really practical stuff," he says. "All this really hard-core work we'd done on how to represent knowledge about concepts like fluids and countries and people." But the conference organizers confined the Cyc papers and speakers to a single panel, Lenat recalls, and few people showed up. "It was like they'd rather talk about doing things than actually hear from people who are doing things," he complains.

Whatever the theoretical inelegance of Cyc, Lenat can always fall back on one powerful defense: it works well — or at least better than anything comparable so far. During a visit to Cycorp, I watch a videotaped demonstration of a biological warfare expert teaching Cyc about a new weapon similar to anthrax. Cyc demonstrates its grasp of common knowledge — not just about the physical world, but about the rather more ephemeral world of pop culture.

At one point, the expert asks Cyc what it knows about anthrax. Cyc pauses for a second, then asks: "Do you mean Anthrax (the heavy metal band), anthrax (the bacterium), or anthrax infection (the infection)?" The official notes that it's the bacterium, not the band. Cyc asks what type of organism the agent infects. The military official types: "People."

Cyc thinks again for an instant, then responds: "By People, I assume you mean persons, not *People* magazine."

By the end of the exchange, Cyc has successfully absorbed various facts about the agent — how it is destroyed (by encasing it in concrete), its color (green), and which terrorists possess it (Osama bin Laden). But the demo also illustrates some of Cyc's limitations. Obviously, it wasn't easily able to figure out the military context of the exchange — otherwise it wouldn't have needed to ask whether "anthrax" signified a heavy metal band.

An even bigger limitation lurks behind that one: the fact that common sense is almost infinite. This is a problem that still threatens to doom Lenat's grandest ambitions. Sure, you could eventually input the billions of bits of common knowledge worldwide. But at the rate the Cyclists are going, that would take millennia; the limited resources of Cycorp's programmers are not enough. Even Cyc's supporters see this as a major stumbling point. "The amount of knowledge you need will easily outpace the ability of the builders to input it," as Nilsson says.

Part of the problem, Lenat concedes, has been that nobody except for scientists and philosophers trained in formal-logic languages can master CycL well enough to input knowledge reliably into Cyc. "This isn't a skill the average person has," he says. Moreover, Cyc's early development was particularly sensitive. Because one small piece of wrong information could cause enormous problems later, Lenat had to be careful, he says, to prevent erroneous material from getting in. Even though he gave out restricted parts of Cyc's knowledge base to academics to examine and experiment with, he didn't accept new knowledge from them.

But Lenat remains optimistic. As Cyc grows larger and more robust, he says, it is becoming less fragile and more likely to detect nonsense inputs. Starting this fall, Lenat will release parts of Cyc — for free — to anyone who wants them. Technically sophisticated users will be able to add knowledge to their own copies of Cyc and, if they choose, send the new information back to Cycorp to be integrated into the master copy — the "master" Cyc, as it were. For the first few years, Lenat's team will carefully scan the outside knowledge to make sure it catches any problems or nonsense information that Cyc doesn't flag on its own. But if all goes well, Cyc will be able to harness the worldwide labor of those who want to input new facts — allowing Cyc's knowledge base to grow at a far more rapid clip than ever before.

Several years down the line, when Lenat has sufficiently improved Cyc's understanding of ordinary English, anyone — nonscientists, average Joes, whoever — will be able to talk to it. Eventually Cyc could even be turned loose on the Web, allowing it to read and absorb the mind-boggling collection of information on the Internet. When Cyc becomes an open-source brain, conversant in everyday English, the pace of its growth could be explosive. "The number of people who can help it grow will increase from a few hundred to a billion," Lenat marvels.

But if Cyc becomes open for anyone to talk to, what will it learn? Will people lie, dissemble, or try to delude it? Last fall, MIT's Push Singh assembled a similar experimental on-line project, Open Mind, to collect everyday intelligence. He was a fan of Cyc and wanted to try building his own knowledge database, but he didn't want to spend decades crafting commonsense statements. He met with David Stork, chief scientist at Ricoh Silicon Valley, who has been working on projects that allow multiple people to collaborate on-line. Singh and Stork realized that an open-source approach would be useful for gathering common knowledge and quickly assembled an on-line database at www.openmind.org to solicit facts from nonspecialists. It was up and running by late last summer; by the summer of 2001, nearly seven thousand people had input more than 300,000 facts.

I look over a handful of the entries. Some are obviously uncontroversial, such as "A square is a closed shape with four equal sides at right angles" or "An adult male human is called a man." Others veer into the realm of custom and opinion, such as "The first thing you do when you hang out at a bar is order a drink" and "Christmas is a commercialized holiday."

For now, Singh has no way of checking the information's veracity, other than reading each input himself. "At this point, we're just seeing whether this is a viable way to collect commonsense knowledge," he says. But after examining several thousand of the information pieces that have been input, he's found that contributors are by and large honest. Their mistakes are not so much sneaky as inadvertent, the result of unclear writing. A better trained user, for example, might have modified the example above to say "Many people feel Christmas is a commercialized holiday." As Singh notes: "People really want to be of help! But they're untrained in how to express information clearly."

Though Open Mind is a collection of plain-language statements that does not include a formal logical language, Lenat is watching Singh's project with interest. When he opens up Cyc to the layperson, he'll face the same challenges. Cyc's basic common sense will have to be robust enough to recognize clearly erroneous information and reject it. More problematically, it will have to recognize when it encounters a subjective belief and categorize it as such. For example, Cyc might read some information about a car on a General Motors Web site. Lenat says Cyc ought to trust basic facts on the site, such as the identifying numbers of specific car parts, since "General Motors is actually the best expert on that." But other material — such as "supposedly third-party reports that just happen to favor General Motors," Lenat notes sardonically — ought to be disregarded.

The potential for mistakes is serious: Cyc's accepting a belief as a fact would be akin to an impressionable young kid's absorbing a dubious bit of information from an adult. At that point, the only way to fix it would be to do "brain surgery" — have a Cyclist go in and manually rewrite the fact in CycL. If Cyc is open and there are millions and millions of new facts being input every day, that could be virtually impossible.

"You begin to see just how complex this could be," Lenat says.

*

Still, he's willing to try. After so many years of pounding away at Cyc, Lenat has nothing to lose in pressing ahead. He draws me a graph that shows Cyc's learning curve. From 1985 to 2000, the line curves upward gradually — the "brain surgery" phase during which the Cyclists input knowledge by hand. But then at 2001, the curve steepens dramatically as the open-source phase takes over, and thousands — or millions — more inputters join in. Lenat extends the curve maybe ten years into the future. As the curve reaches the point where Cyc has read everything there is to read and spoken with everyone willing to tell it facts, it will begin to flatten out. "It'll know all there is to know," he says. "At that point, the only way it could learn more is by doing experiments itself."

Will 2001, then, be as talismanic as some would hope — the year that HAL-like intelligence is born? Lenat is optimistic. "I'm very, very excited," he says. But he's made rash predictions in the past: in the late 1980s, he confidently forecast that 1994 would be the year Cyc would begin learning via natural-language conversations with average people.

Intelligence is unruly stuff — which makes the behavior of artificial intelligence sometimes hard to predict. Lenat tells me a cautionary tale from his days as a young professor in the late 1970s and early 1980s. Back then, he designed a self-learning program called EURISKO. It was intended to generate new heuristics — new types of search strategies — all on its own by slightly mutating bits of LISP computer code. And it did successfully manage to produce unique new rules to parse data sets.

But then trouble struck. In the mornings when Lenat arrived at work, he'd find the program had mysteriously shut down overnight. This happened again and again, puzzling him. What was causing it to crash?

Finally he discovered that it wasn't a crash at all; it was a strange and unexpected new strategy. At some point, EURISKO had altered its rules so that "making no errors at all" was as important as "making productive new discoveries." Then EURISKO realized that if it turned itself off, it wouldn't make mistakes. It was bizarre behavior — but it was logical. It made sense. "Just not common sense," Lenat says with a laugh.

JOY WILLIAMS

One Acre

FROM *Harper's Magazine*

I HAD AN ACRE in Florida, on a lagoon close by the Gulf of Mexico.

I am admittedly putting this first line up against Isak Dinesen's famous oneiric one: *I had a farm in Africa, at the foot of the Ngong Hills.* When Dinesen first came to Africa she confessed that she could not "live without getting a fine specimen of each single kind of African game." For her the hunt was an eroticized image of desire, "a love affair," wherein the "shot . . . was in reality a declaration of love." She must have blushed to read this drivel later, for after ten years she found hunting "an unreasonable thing, indeed in itself ugly and vulgar, for the sake of a few hours enjoyment to put out a life that belonged in the great landscape and had grown up on it." One could say her thinking had evolved, that she had become more conscientious. Still, when she was about to leave her beloved farm (her house, empty of furniture, was admirably "clean like a skull"), she planned to shoot her dogs and horses, dissuaded from doing so only by the pleas of her friends. The animals belonged to her, as had the land, which she ceased to own only when it became owned by another, and subject to that person's whims and policies. Of course it became hers again through writing about it, preserving it in *Out of Africa.* Once again, Art, reflective poesy, saves landscape.

I had an acre in Florida . . . This bodes no drama. For what wonders could a single acre hold, what meaning or relevance? Although the word "Florida" is oneiric, too, and thus its own metaphor. It is an occasional place, a palmed and pleasant stage for transients. To hold fast to an acre in that vast state is almost neurotic. An acre is both too much and not enough. Its value lies in its divisibility, in how many building lots are permitted by law. Four, certainly.

I once saw a white heron in a tumbled landscape on the sprawling outskirts of Naples, a city that crowds against the Big Cypress National Preserve and Everglades National Park. The heron seemed to be beating its head against a tree knocked down by bulldozers to widen a road. Water still lay along the palmetto-dotted earth, but pipes would soon carry it away and dry the land for town houses and golf courses. Cars sped past. The heron, white as a robed angel must surely be, was beating his head against the tree. He was lost to himself, deranged, in his ruined and lost landscape.

I have seen all manner of beautiful waterbirds struck down by cars. I used to take them home and bury them between the mangroves and the live oaks on my lagoon. But of course it was not my lagoon, this body of water only a mile and a half long. To the north it cedes to a private road that gives access to the Sanderling Club, where the exceptionally wealthy enjoy their Gulf views. To the south it vanishes beneath the parking lot for a public beach. This is on Siesta Key, a crowded eight-mile-long island off Sarasota that is joined to the mainland by two bridges, one four lanes, the other two. The lagoon is named Heron; the beach, Turtle. Yes, the turtles still come to nest, and the volunteers who stake and guard the nests are grateful — they practically weep with gratitude — when the condo dwellers keep their lights out during the hatching weeks so as not to confuse the infant turtles in their night search for the softly luminous sea. But usually the condo dwellers don't keep their lights out. They might accommodate the request were they there, but they are seldom there. The lights are controlled by timers and burn bright and long. The condos are investments, mostly, not homes. Like the lands they've consumed, they're cold commodities. When land is developed, it ceases being land. It becomes covered, sealed, its own grave.

Ecosystems are something large to be saved, if at all, by the government at great expense and set aside to be enjoyed by all of us in some recreational or contemplative fashion. An individual doesn't think of himself as owning an ecosystem. The responsibility! Too much. Besides, there's something about the word that denotes the impossibility of ownership. *Land,* on the other hand, is like a car or a house; it has economic currency. Aldo Leopold defined land as *a fountain of energy flowing through a circuit of soils, plants, and animals;* it was synonymous with ecosystem, and he argued that we all have an obligation to protect and preserve it. It was over fifty years ago that Leopold wrote his elegantly reasoned essay "The Land Ethic," but it has had about as much effect on the American conscience as a snowflake. Seven thousand acres are lost to development each day in this country. Ecosystem becomes land becomes parcel.

On Siesta Key, "open space" (of which there is now none), when bought by the county years ago, is being utilized as beach parking or tennis courts. "Raw" land no longer exists, though a few lots are still available, some with very nice trees, most of which will have to go (unfortunately) in order to accommodate the house that will be built on what is now considered a "site." We hardly can get all ecosystem emotional over a site. A banyan tree will most assuredly have to go because it is in its nature to grow extravagantly and demand a great deal of space. Trees, of course, cannot *demand* anything. As with the wild animals who have certain requirements or preferences — a clutter and cover, long natural hours of friendly concealing dark — anything they need can be ignored or removed right along with them.

In 1969, I bought Lots 27, 28, and 29 on Midnight Pass Road, a two-lane road that ended when the key did. There was a small cypress house, no beauty, and an even smaller cypress cottage. They were single-story affairs with flat roofs, built on poured slabs. The lots together cost $24,000. In 1972, I bought Lots 30 and 31 for $12,000. Lagoon land wasn't all that desirable. There was no access to open water. Bay land was more valuable, and even then Gulf front was only for the wealthy. Beachfront is invaluable because no one can build on the sea; the view is "protected." I could hear the Gulf on my small acre; it was, in fact, only several hundred feet away, concealed by a scrim of mangroves. The houses that were to be built over there were grand but still never quite exceeded the height of the mangroves. I did not see my lagoon neighbors for my trees, my tangled careless land, though as the years passed I put up sections of wooden fence, for my neighbors changed, then changed again, and their little cypress houses were torn down in a twinkling, the "extra" lots sold. I put a wooden fence up along the road eventually. It weathered prettily but would shudder on its posts from a flung beer bottle. Sections of it were periodically demolished by errant cars. I don't believe I ever rushed sympathetically to the befuddled driver's aid. No one actually died, but they did go on, those crashes. Streetlights went up at fifty-foot intervals on the dark and curvy road. The bay side got the lights, the lagoon side got the bicycle path. Homeowners were responsible for keeping the "path" tidy, and I appeared out there dutifully with broom and rake, pushing away the small oak leaves from the trees that towered overhead, disclosing all that efficient concrete for the benefit of increasing streams of walkers and joggers. Bicyclists preferred to use the road. Any stubborn palmetto that fanned outward or seeded palm that once graced the strip of land outside my rickety wall would be snipped back by a supernumerary, doing his/her part for the public way. The bottles, cans, and wee chip bags were left for me to reap. As owner of Lots 27, 28, 29, 30, and 31, I had 370 feet of path to maintain. I became aware, outside my fence, of the well-known Florida light, a sort of blandly insistent *urban* light — feathery and bemused and resigned. Cars sped past. Large houses were being constructed on the bay, estates on half an acre with elaborate wrought-iron fences and electric gates. Palmetto scrub had given way to lawns. Trees existed as dramatically trimmed accents, all dead wood removed. Trees not deemed perfectly sound by landscape professionals were felled; the palms favored were "specimen" ones. Dead animals and birds appeared more and more frequently on the road. Cars sped past.

Behind my wobbly fence, I pottered about. The houses were built in the forties, and the land had the typical homesteaded accoutrements of that time — a few citrus trees, some oleander and hibiscus for color, a plot cleared for a few vegetables and Shasta daisies, a fig left to flourish for shade, and live oaks left to grow around the edges. The ghastly malleucas were available in nurseries then and were often planted in rows as a hedge. The man I bought the land from was a retired botanist, and he had planted avocado and lychee nut trees, too, as well as a grove of giant bamboo from which he liked to make vases and bowls and various trinkets. There was bougainvillea, azalea, gardenia, powder puff and firecracker plant, crotons, wild lilies, sea grape, and several orchid trees. Of the palms there was a royal, sabal, many cabbage, pineapple, sago — queen and king — reclinata, fishtail, sentry, traveler's, and queen. There were cypresses, jacaranda, and two banyan trees. There was even a tiny lawn with small cement squares to place the lawn chairs on. The mangroves in this spot had been cut back for a view of the idly flowing tea-colored lagoon. Elsewhere they grew — the red and the black — in the manner each found lovely, in hoops and stands, creating bowers and thickets and mazes of rocking water and dappling light.

This was my acre in Florida. Visitors ventured that it looked as though it would require an awful lot of maintenance, though they admired my prescience in buying the extra lots, which would surely be worth something someday. The house had a certain "rustic charm," but most people didn't find the un-air-conditioned, un-dehumidified air all that wholesome and wondered why I kept the place so damn dark, for there were colored floods widely available that would dramatize the "plantings." I could bounce more lights off the water, you could hardly even tell there was water out there, and what was the sense of hiding that? And despite the extraordinary variety, my land seemed unkempt. There were vines and Brazilian pepper and carrotwood, there were fire-ant mounds, rats surely lived in the fronds of the untrimmed palms. My acre looked a little hesitant, small and vulnerable, *young.* Even the banyan tree was relatively young. It had put down a few aerials but then stopped for a good decade as if it were thinking . . . *What's the use. I'm straddling Lots 29 & 30 and I'm not known as an accommodating tree. When the land gets sold, I'll be sold, too, and will fall in screaming suttee* . . . Or sentiments of that sort.

As for the birds and animals, well, people didn't want raccoons and opossums and armadillos, and their cats would eat the baby rabbits. Too disgusting, but that's just the way nature *was.* And although I had cardinals and towhees and thrashers and mockingbirds and doves and woodpeckers, they did, too; as a matter of fact their cardinals were nesting in a place where they could actually see them, right near the front door, and that was getting to be quite the nuisance. As for the herons, you found them everywhere, even atop the dumpsters behind the 7-Eleven. Such beggars. You had to chase them away from your bait bucket when you were fishing from the beach. Did I fish in the lagoon? There were snapper in there, redfish, maybe even snook. I could get the mullet with nets. Why didn't I fish?

The years flowed past. Some of the properties on the lagoon fell to pure speculation. Mangroves were pruned like any hedging material; in some cases, decks were built over them, causing them to die, though they remained ghostily rooted. Landowners on the Gulf did not molest their mangrove. The lagoon to them was the equivalent of a back alley. Why would they want to regard the increasing myriad of houses huddled there? I traveled, I rented the place out, I returned. There were freezes, we were grazed by hurricanes. An immense mahoe hibiscus died back in a cold snap, and two years later a tall, slender, smooth-barked tree it had been concealing began producing hundreds of the pinkest, sweetest, juiciest grapefruit I have ever tasted. The water oaks that had reached their twenty-year limit rotted and fell. There were lovely woodpeckers. All through the winter in the nights the chuck-will's-widows would call.

That would drive me nuts, several of my acquaintances remarked.

The sound of construction was almost constant, but no one appeared to be actually living in the remodeled, enlarged, upgraded properties around me. I had cut out sections of my side fences to allow oak limbs to grow in their tortured specific manner, but my neighbor's yardmen would eventually be instructed to lop them off at the property line. This was, of course, the owner's right. There was the sound of trimmers, leaf-blowers, pool pumps, pressure cleaners; the smell of chemicals from pest and lawn services. Maintenance maintenance maintenance. Then the county began cutting back the live oak limbs that extended over the bicycle path, even though one would have to be an idiot on a pogo stick to bump into one. Sliced sure as bread, the limbs, one at least five feet in diameter and green with resurrection fern and air plants, were cut back to the fence line.

It was then I decided to build the wall.

The year was 1990. The wall was of cement block with deep footers, and it ran the entire length of the property except for a twelve-foot opening, which was gated in cedar. It cost about $10,000, and two men did it in two months. The wall was ten feet high. It was not stuccoed. I thought it was splendid. I didn't know many people in the neighborhood by then, but word got back to me that some did not find it attractive. What did I have back there, a prison? To me, it was the people speeding past the baby Tajs on the animal-corpse-littered road who had become imprisoned. Inside was land — a mysterious, messy fountain of energy; outside was something else — not land in any meaningful sense but a diced bright salad of colorful real estate, pods of investment, its value now shrilly, sterilely economic.

Behind the wall was an Edenic acre, still known to the tax collector as Lots 27, 28, 29, 30, and 31. Untransformed by me, who was neither gardener nor crafty ecological restorer, the land had found its own rich dynamic. Behind the wall were neither grounds nor yard nor garden, nor park, nor even false jungle, but a functioning wild landscape that became more remarkable each year. Of course there was the humble house and even humbler cottage, which appeared less and less important to me in the larger order of things. They were shelters, pleasant enough but primarily places from which to look *out* at the beauty of a world to which I was irrelevant except for my role of preserving it, a world I could be integrated with only to the extent of my not harming it. The wildlife could hardly know that their world in that place existed only because I, rather than another, *owned* it. I knew, though, and the irrationality of the arrangement, the premise, angered me and made me feel powerless, for I did not feel that the land was mine at all but rather belonged to something larger that was being threatened by something absurdly small, the ill works and delusions of — as William Burroughs liked to say — *Homo sap.*

Although the wall did not receive social approbation, its approval from an ecological point of view was resounding. The banyan, as though reassured by the audacious wall, flung down dozens of aerial roots. The understory flourished; the oaks soared, creating a great grave canopy. Plantings that had seemed tentative when I had bought them from botanical gardens years before took hold. The leaves and bark crumble built up, the ferns spread. It was odd. I fancied that I had made an inside for the outside to be safe in. From within, the wall vanished; green growth pressed against it, staining it naturally brown and green and black. It muffled the sound and heat of the road. Inside was cool and dappled, hymned with birdsong. There were owls and wood ducks. An osprey roosted each night in a casuarina that leaned out over the lagoon, a tree of no good reputation and half-dead, but the osprey deeply favored it, folding himself into it invisibly in a few seconds each nightfall. A pair of yellow-crowned night herons nested in a slash pine in the center of Lot 30. Large birds with a large hidden nest, their young — each year three! — not hasty in their departure. A single acre was able to nurture so many lives, including mine. Its existence gave me great happiness.

And yet it was all an illusion, too, a shadow box, for when I opened the gates or canoed the lagoon, I saw an utterly different world. This was a world that had fallen only in part to consortiums of developers; it had fallen mostly plat by plat to individuals, who, paradoxically, were quite conformist in their attitude toward land, or rather the scraped scaffolding upon which their real property was built. They lived in penury of a very special sort, but that was only my opinion. In *their* opinion they were living in perfect accord with the values of the time, successfully and cleverly, taking advantage of their advantages. Their attitudes were perfectly acceptable; they were not behaving unwisely or without foresight. They had maximized profits, and if little of nature had been preserved in the arrangement, well, nature was an adornment not to everyone's taste, a matter really of personal tolerance and sympathy. Besides, Nature was not far away, supported by everyone's tax dollars and preserved in state and federal parks. And one could show one's appreciation for these places by visiting them at any time. Public lands can be projected as having as many recreational, aesthetic, or environmental benefits as can be devised for them, but private land, on this skinny Florida key and almost everywhere in this country, is considered too economically valuable to be conserved. Despoliation of land in its many, many guises is the custom of the country. Privately, one by one, the landowner makes decisions that render land, in any other than financial terms, moot. Land is something to be "built out."

In contrast to its surrounds, my acre appeared an evasion of reality, a construct, a moment poised before an inevitable after. How lovely it was, how fortunate I was. Each day my heart recognized its great worth. It was invaluable to me. The moment came when I had to sell it.

*

Leopold speaks of the necessity of developing an ecological conscience, of having an awareness of land in a philosophical rather than an economic sense. His articulation of our ethical obligations to the land is considered by many to be quite admirable. We celebrated the fiftieth anniversary of this articulation (if not its implementation) in 1999. A pretty thought, high-minded. And yet when one has to *move on* (if not exactly in the final sense) one is expected to be sensible, realistic, even canny, about property. I was not in the comfortable kind of financial situation where I could deed my land to a conservation group or land trust. Even if I could have, it would probably have been sold to protect more considerable sanctuary acreage elsewhere, for it was a mere acre in a pricey neighborhood, not contiguous with additional habitat land, though the lagoon did provide a natural larger dimension. I had been developing an ecological conscience for thirty years, and I could continue to develop it still certainly, become a good steward somewhere else, because once I had decided to sell, this particular piece of land and all the creatures that found it to be a perfect earthly home would be subject to erasure in any meaningful ecological sense, and this would not be considered by society to be selfish, cruel, or irresponsible.

"Wow, it's great back here," the realtor said. "I often wondered what the heck was going on back here. I'm looking forward to showing this place."

I told him I wanted to sell the land as a single piece, with deed restrictions, these being that the land could never be subdivided; that the buildings be restricted to one house and cottage taking up no more land than the originals; and that the southern half of the property be left in its natural state as wildlife habitat.

"Nobody wants to be told what they can do with their land," he told me, frowning. "I'll mention your wishes, but you'll have to accept a significant reduction in price with those kinds of restrictions. When we get an offer you and the buyer can negotiate the wording of the agreement. I'm sure the type of person who would be attracted to this property wouldn't want to tear it all apart."

"Really," I said, "you don't think?"

I went through a number of realtors.

With a lawyer I drew up a simple and enforceable document that the realtors found so unnerving that they wouldn't show it right away to interested parties, preferring word wobble and expressions of goodwill. There were many people who *loved* the land, who *loved* nature, but would never buy anything that was in essence not free and clear. Or they had no problem with the restrictions *personally,* but when they had to sell (and Heaven forbid that they would right away of course) they could not impose such coercive restraints on others. The speculators and builders had been dismissed from the beginning. These were people of a more maverick bent, *caring* people who loved Florida, loved the key — wasn't it a shame there was so much development, so much change. When they saw the humble document they said 1) who does she think she *is*? 2) she's crazy, 3) she'll never sell it. Over the months the realtors took on a counseling manner with me as though I needed guidance through this dark stubborn wood of my own making, as though I needed to be talked down from my irrational fanciful resolve. They could sell the land for $200,000 more if I dropped the restrictions. My acre could be destroyed naturally; a hurricane could level everything, and the creatures, the birds, would have to go somewhere else anyway. With the money I'd make marketing it smartly, I could buy a hundred acres, maybe more, east of the interstate. There was a lot of pretty ranchland over there. I could conserve that. A lot of pressure would be on that land in a few years; I could do more by saving that. Sell and don't look back! That's what people did. You can't look back.

I'm not looking back, I said.

And I wasn't.

I was looking ahead, seeing the land behind the wall still existent, still supporting its nests and burrows — a living whole. I was leaving it — soon I would no longer be personally experiencing its loveliness — but I would not abandon it, I would despise myself if I did. If I were to be party to a normal real estate transaction, I would be dooming it, I would be — and this is not at all exaggerated — signing a warrant for its death. (Perhaps the owners of the four new houses that could — and would, most likely — be built would have the kindness to put out some birdseed.) I wanted more than money for my land, more than the mere memory of it, the luxury of conserving it falsely and sentimentally through lyrical recall. I wanted it to *be.*

It took eight months to find the right buyer. Leopold's "philosophers" were in short supply in the world of Florida real estate. But the ideal new owners eventually appeared, and they had no problem with the contract between themselves and the land. I had changed no hearts or minds by my attitude or actions; I had simply found — or my baffled but determined realtors had — people of my persuasion, people who had a land ethic, too. Their duties as stewards were not onerous to them. They did not consider the additional legal documents they were obliged to sign an insult to their personal freedom. They were aware that the principle was hardly radical. An aunt had done a similar thing in New England, preserving forty acres of meadow and woodland by conservation easement. They had friends in California who had similarly sold and conserved by deed four hundred acres of high desert. And here was this enchanted acre.

It had been accomplished. I had persisted. I was well pleased with myself. Selfishly I had affected the land beyond my tenure. I had gotten my way.

And with all of this, I am still allowed to miss it so.

KAREN WRIGHT

Very Dark Energy

FROM *Discover*

IN JANUARY OF 1917, Albert Einstein was putting the finishing touches on his general theory of relativity when he decided to cheat just a little. The man who said that imagination is more important than knowledge was trying to use his new theory to solve an old puzzle of the cosmos, and he wasn't getting anywhere. Under Newton's laws, stars and other heavenly bodies pull on one another through the force of gravity. A countervailing propulsion, like a big explosion, could overcome that attraction, but once it fizzled out, gravity would start pulling things together again. Either way, matter in the universe should be moving — either hustling out into space or clumping into a kind of cosmic hairball.

But the universe that Newton and Einstein knew was a tame, stable place. The Milky Way was the only galaxy in town, and its stars seemed fixed in the firmament. The seeming stasis of the night sky had stumped Newton, and even a theory as powerful as relativity failed to explain it. So Einstein added an arbitrary term to his equations. Mathematically, it acted like a repulsive force spread smoothly throughout the universe. Where gravity pulled, he said, this force pushed back in equal measure. He called this fudge factor lambda, and eventually it came to be known as the cosmological constant.

Einstein never felt good about lambda, because he couldn't point to any theoretical or experimental evidence for its existence. Later in life he called it his greatest blunder. "Admittedly," he wrote, "[lambda] was not justified by our actual knowledge of gravitation." But Einstein's imagination was always more powerful than the knowledge of his day, and now, nearly a century later, his blunder is beginning to look like yet another stroke of uncanny genius.

In the last seventy-five years, astronomers have radically revised their conception of the cosmos. Edwin Hubble showed in 1929 that the universe was not static but expanding — it was getting bigger all the time, as if some primal explosion were driving its contents apart. That primal explosion came to be known as the Big Bang, and the expanding universe was its love child. For fifty years, Big Bang cosmology reigned.

Then, three years ago, light from distant, dying stars revealed that the edges of space are rushing away from one another at an ever-increasing rate. The cosmos, it seems, is not just growing but growing faster and faster. The bigger the universe gets, the faster it grows. Some ubiquitous, repulsive force is driving at the margins of space, stomping on the accelerator. And there are no red lights in sight. That mysterious propulsion looks a lot like lambda.

Today's cosmologists are calling this force dark energy: "dark" because it may be impossible to detect, and "energy" because it's not matter, which is the only other option. Despite the sinister connotations, dark energy is a beacon that may lead physicists to an elusive "final theory": the unification of all known forces, from those that hold the components of atoms together to the gravity that shapes space. Meanwhile the notion of dark energy has helped reconcile a puzzling suite of recent observations about the shape and composition of the cosmos.

In fact, the future of physics and the fate of the universe may ultimately depend on a kind of antigravity that has heretofore been a subject of mere conjecture. The experts think they know what role dark energy plays in the cosmos. Now all they have to do is figure out what dark energy is.

Hubble and his fellow astronomers discovered the expansion of the universe by observing that galaxies in all directions are moving farther away from one another all the time. He was able to track this movement through a phenomenon called redshift, in which visible starlight gets stretched out into longer wavelengths (toward the red end of the visible light spectrum) as it moves through expanding space. The amount of redshift depends on the rate of cosmic expansion and the distance of the observer from the galaxy.

Einstein, Newton, and most other physicists had assumed that gravity would put the brakes on expansion. But decades after Hubble's breakthrough, astronomers were still trying to measure the assumed deceleration. The answer finally came in the late 1990s, from giant telescopes studying the light of stars dying in spectacular explosions called supernovas. Supernovas are among the brightest events in the cosmos, so they can be seen from very far away. Because light from the most distant supernovas must travel for billions of years to reach our telescopes, astronomers can look to its redshift for a historical record of expansion reaching back billions of years.

At a meeting in Washington, D.C., three years ago, a team of researchers from the Lawrence Berkeley Laboratory showed that the light from very distant supernovas is stretched out less than was predicted given the current rate of expansion. Apparently, the universe expanded more slowly in the past than it does now. Expansion isn't slowing down as expected; it's speeding up. The finding was counterintuitive, and it was based on brand-new methodology. But at the same time, a second group of space-telescope studies led by Brian Schmidt and Robert Kirshner of the Harvard-Smithsonian Center for Astrophysics came to the same conclusion.

"It seemed like we must have done something wrong," says Kirshner. "The cosmological constant had such a bad stink, you know? I mean, 'Einstein screwed up. What makes you think you're gonna do any better?'"

"I was floored," echoes University of Chicago cosmologist Michael Turner, recalling his first encounter with the evidence at the Washington meeting. "Yet everything fell into place. This was the answer we'd been looking for."

In particular, Turner was looking for a way to resolve conflicting results that were turning up in other experiments describing the state of the cosmos. One set of studies sought to determine the shape of the universe by considering the density of matter in it. Einstein had shown that matter curves space in predictable ways, so that universes with different densities of matter will have different shapes. His theories allowed for three shapes: negative curvature, in which the universe looks like a saddle; positive curvature, in which the universe is spherical; and flat, the most unlikely case, in which the overall density of matter doesn't warp space, and photons travel in straight lines. Flat space isn't two-dimensional; it just isn't curved.

Each shape corresponds to a density of matter denoted by the symbol omega. To create a flat universe, matter must reach so-called critical density, which means omega equals one. In a saddle-shaped universe, omega is less than one; in a spherical universe, it's more than one. Astronomers have sought to determine the value of omega and distinguish among these geometries by measuring the way space bends beams of light. The light they like to measure isn't visible; it's microwave radiation left over from the Big Bang that glows at the farthest reaches of the universe. Distortions in that microwave signal can reveal the shape of the intervening space. In a saddle-shaped universe, distinct patches of the microwave background would look smaller than they're predicted to be. A sphere-shaped universe would magnify the patches of background radiation. In a flat universe, patches of background radiation would be closest to their predicted size.

Recent studies of microwave background radiation had hinted that the universe is flat. But last spring, data from balloon-borne instruments lofted over Texas and Antarctica supplied convincing evidence. Minute fluctuations in the radiation were the expected size. The most precise measurements available revealed that the shape of the universe is flat; it has the critical density and omega equals one.

Unfortunately, these findings don't match results from inventories of matter in the universe. The density of matter can be inferred from its regional gravitational effects on light and on the evolution of galaxies. When astronomers use these methods to tally up the contents of the cosmos, all the people, planets, galaxies, and gases put together account for less than a tenth of the density predicted by the microwave background data. Even the most exhaustive surveys, which include exotic forms of matter only recently discerned, find just a third of the critical density. There's not nearly enough stuff to account for the flatness astronomers observe. Unlikely as it seems, says Turner, the universe seems to be made up mostly of empty space — a vacuum.

"And that finding," says University of Texas physicist Steven Weinberg, "could be regarded as the most fundamental discovery of astronomy."

Weinberg is a Nobel Prize–winning particle physicist who has spent most of his life describing theoretical forms of energy that haven't been discovered yet. The discrepancy between the microwave background and the matter surveys intrigued him, because he knew that energy can shape space just as matter does. A flat universe, or indeed a universe of any shape, could well be molded by both matter and energy. Einstein had recognized this possibility when he perceived that energy and matter are essentially equivalent — as in $E = mc^2$. Thus, he knew that energy could constitute the missing two-thirds of the critical density.

And unlike Einstein, Weinberg and his fellow theorists had never quite given up on the old idea of the cosmological constant — some widespread energy loitering in empty space. As quantum mechanics matured through the middle of the last century, it began to make sense, in a wonky way, that the apparent vacuum might have some energy in it. Theorists had even named the hypothetical vacuum energy lambda, in honor of Einstein's goof. And they'd realized long ago that if energy in the vacuum exists, it has a repulsive effect — one that could cause a universe to accelerate.

But if some exotic form of repulsive energy does make up two-thirds of all the stuff in the universe, it must be very weak. Otherwise its effects would have been obvious long ago. Whatever the mysterious lambda is, it must do its work only across great distances, on a cosmic scale.

That was the nature of Turner's epiphany in Washington three years ago. The light from remote supernovas showed that some unknown repulsive force was speeding up the expansion of the universe. And the microwave data and the matter surveys only made sense if such a force existed. All the evidence pointed to the presence of a kind of energy that so far had existed only on paper. As he was standing in front of a poster from the Lawrence Berkeley lab, Turner put all of the puzzling pieces together.

"The discovery of an accelerating universe was simultaneously the biggest surprise and the most anticipated discovery in astronomy," he says. It put dark energy on the map.

So the universe circa 2001 is flat, accelerating, and very nearly empty. And astronomers are happy, because a single entity with Einstein's imprimatur can explain all these attributes. But if the existence of dark energy has simplified researchers' understanding of the contemporary cosmos, it has also introduced plenty of complications. One has to do with the fate of the universe.

In the days before dark energy, astronomers believed that the end of the expanding universe would be dictated by the density of matter in it. Just as matter determined the curvature of space, it would also predict the way that space would expand and whether it would ever contract. Back when cosmic expansion was caused solely by the cataclysmic propulsion of the Big Bang, the gravity of matter was expected to eventually slow it down, maybe even stop it, maybe even reverse it. In short, density equaled destiny.

Based on that reasoning, astronomers proposed three models for the fate of the universe, each corresponding to a different geometry and density of matter. In each scenario, the gravitational attraction of all the matter in the universe tugs at the heels of the Big Bang's momentum like a tireless dog that's latched onto the leg of a running mailman.

If omega is less than one, the universe keeps on expanding forever, but at an ever-diminishing pace. That universe has the saddle shape and is called "open." If omega is more than one, the universal expansion slows and eventually reverses, collapsing in a cosmic crunch. That universe is spherical and "closed." In a flat universe, where the density of matter is exactly one, the expansion eventually slows very nearly to a stop but never actually reverses.

But if the universe is made up mostly of repulsive, ubiquitous energy rather than matter, then its ultimate fate isn't inscribed in its shape after all.

"We used to say that fate and geometry were connected," says Turner. "But that's only true if the stuff of the universe is matter alone. Once dark energy comes in, then destiny and geometry decouple. So you can have a closed universe that expands forever and an open universe or a flat universe that collapses."

The only way to figure out the fate of the flat, empty, accelerating universe, says Turner, is to learn more about the dark energy that's impelling expansion. But even as they begin chasing down Einstein's notion of vacuum energy, physicists are having to grapple with problems that range from the numerical to the philosophical. For one thing, when they attempt to calculate the value of lambda, the theorists come up with a figure that is 120 orders of magnitude too big. Not 120 times too big — 10^{120} times too big. Fitting the known universe with a vacuum energy of that potency would be like filling up a water balloon with a fire hose.

"It cannot possibly be correct," says Turner. "If it were correct, you wouldn't be able to see beyond the end of your nose, the universe would be expanding so fast." The size of the error has emphasized how poorly physicists understand certain aspects of gravity. "That is the biggest embarrassment in theoretical physics," adds Turner.

It gets even more embarrassing, because theorists can't explain why the densities of matter and energy are currently so close in value. Theoretically, either of those densities could be anything from zero to infinity, and their ratio could vary accordingly. The odds of their being within an order of magnitude of each other are very low. The precarious balance between matter and energy that exists today in our universe — one-third matter to two-thirds energy — seems as improbable as the static universe that Einstein struggled to describe. And some find that improbability especially suspicious, because a universe more dominated by dark energy would be inhospitable to life. The excess energy would prevent matter from clumping into galaxies, stars, and planets. Yet here we are.

The coincidence has driven even notorious skeptics like Weinberg to invoke, in exasperation, the anthropic principle. That much-maligned tautology states that human consciousness can question the terms required for its existence only in a world in which those terms have been met. If conditions were any different, no one would be here to ponder them.

"I don't like this kind of argument," Weinberg admits. "But I don't know of any other explanation that comes close."

The anthropic principle is anathema to most physicists. Some would rather propose a brand-new force in the cosmos than fall back on rhetorical sleight-of-hand. Paul Steinhardt of Princeton University, for example, has already ditched the cosmological constant in favor of a new category of dark energy that he calls quintessence. The fact that energy and matter have achieved a delicate balance is suspicious, he says, only if you assume there's no communication between the two. Steinhardt has proposed that repulsive energy senses the presence of matter and changes its strength or distribution to maintain a balance of densities. This energy could alter its properties over space and across time; unlike lambda, it wouldn't be distributed evenly, and it wouldn't remain constant.

"There was always logically the possibility of having such fields," Steinhardt maintains. "But there was no reason to invoke them, because they weren't required by any theory."

Now that there is, Steinhardt is hoping experimental physicists will turn up evidence of quintessence in minute fluctuations of temperatures in the cosmic microwave background. The MAP satellite scheduled for launch in June could be instrumental in detecting such signals. More detailed surveys of distant supernovas are also planned.

"Different dark-energy models will make different predictions about the evolution of the acceleration of the universe over time," says Saul Perlmutter, leader of the Lawrence Berkeley team. Perlmutter is championing a plan to study acceleration with a space-based telescope called the SuperNova/Acceleration Probe, or SNAP. "We want to go back in history and find out when the universe went through its growth spurts."

Reckoning with dark energy will also spur attempts to define a quantum theory of gravity. Gravity is the only one of the four known forces that has eluded description in terms of energy bundles called quanta. Physicists have already managed to bring the other three — the strong force, the electromagnetic force, and the weak force — into the quantum fold. But unlike those three forces, gravity typically operates on vastly different scales than quantum mechanics. "Gravitation governs the motions of planets and stars," Weinberg wrote in a recent review, "but it is too weak to matter much in atoms, while quantum mechanics, though essential in understanding the behavior of electrons in atoms, has negligible effects on the motions of stars or planets."

With the discovery of dark energy, the two worlds collide. In the acceleration of the universe may lie some clues to the behavior of tiny quanta of gravitational energy. Einstein's own theories of gravity allow it to have some sort of repulsive effect, so elucidating the nature of dark energy could hasten theorists on their way to a final theory unifying all the forces. That's why physicists scanning the furthest reaches of space with powerful telescopes suddenly seem very interesting to the physicists scribbling on the blackboards.

"It's very flattering for astronomers," says Harvard's Kirshner. "We have the attention of the high priests of our field."

But there's no guarantee that dark energy will serve up the eternal verities the high priests are hoping for. The unlikely balance of energy and matter and the strength of the vacuum energy may permit human existence through caprice, not necessity. Einstein himself knew well the hazards of counting on capricious nature. "Marriage," he once opined, "is the unsuccessful attempt to make something lasting out of an accident." Scientists who would seek permanent truths in the accelerating universe could be making the same mistake.

Contributors' Notes

Roy F. Baumeister received his Ph.D. in social psychology from Princeton University in 1978. He currently holds the E. B. Smith Professorship in the Liberal Arts at Case Western Reserve University. As of January 2003, he will be Eppes Professor of Psychology at Florida State University. His fifteen books include *Evil: Inside Human Violence and Cruelty* and *The Social Dimension of Sex.*

Burkhard Bilger is a staff writer for *The New Yorker* and a senior editor at *Discover.* Until 1999, he was a writer and deputy editor for *The Sciences,* where his work helped garner two National Magazine Awards and six nominations. His first book, *Noodling for Flatheads,* was published in October 2000 and was a finalist for the PEN/Faulkner Award for first nonfiction. Bilger lives in Brooklyn with his wife, Jennifer Nelson, their children, Hans, Ruby, and Evangeline, and their coonhound, Hattie.

K. C. Cole writes the "Mind Over Matter" column for the *Los Angeles Times,* where she has covered physical science since 1994, and does a bimonthly science commentary for Pasadena public radio (KPCC). She is the author of *The Hole in the Universe: How Scientists Peered Over the Edge of Emptiness and Found Everything, The Universe and the Teacup: The Mathematics of Truth and Beauty,* and *First You Build a Cloud: Reflections on Physics as a Way of Life.*

Richard Conniff is the author of *Spineless Wonders: Strange Tales from the Invertebrate World* and *The Natural History of the Rich: A Field Guide,* among other books. He is a past winner of the National Magazine Award and an Emmy Award nominee for his work in natural history television.

Frederick C. Crews taught English at the University of California, Berkeley, from 1958 until his retirement in 1994. His works include the best-selling satire *The Pooh Perplex;* critical studies of Henry James, E. M. Forster, and Nathaniel Hawthorne; two widely adopted composition handbooks; two collections of his own essays on psychoanalysis and other matters; a book about American literary criticism, *The Critics Bear It Away* (nominated for the National Book Critics Circle Award), and the recent satire *Postmodern Pooh.*

A social activist and a feminist, **Barbara Ehrenreich** has written on the subjects of health care, class, families, and sex. Among her most respected — and controversial — books are *The American Health Empire: Power, Profits, and Politics, For Her Own Good: 150 Years of the Experts' Advice to Women,* and *The Hearts of Men: American Dreams and the Flight from Commitment.* Originally a biologist who earned her Ph.D. from Rockefeller University, Ehrenreich became involved in political activism during the Vietnam War and has written professionally ever since. Her most recent book is *Nickel and Dimed: On (Not) Getting By in America.*

H. Bruce Franklin is the author or editor of eighteen books and more than two hundred articles on culture and history. He is currently the John Cotton Dana Professor of English and American Studies at Rutgers University in Newark.

Malcolm Gladwell is a writer for *The New Yorker* and the author of *The Tipping Point.* An archive of his work is available at gladwell.com.

Gary Greenberg is a psychotherapist and freelance journalist. He teaches psychology at Connecticut College, and his work has appeared in *Rolling Stone, The New Yorker, Discover,* and *McSweeney's,* among other publications.

Gordon Grice is the author of *The Red Hourglass: Lives of the Predators,* recently listed by the editors of Amazon.com as one of fifteen "Wildlife Classics." His work has appeared in *The New Yorker, Harper's, Discover,* and other magazines. His next book, a memoir, is under contract. He lives in Wisconsin with his wife and their three sons.

Blaine Harden is a national correspondent for the *New York Times* and a writer for the *New York Times Magazine.* Previously, he held numerous positions at the *Washington Post,* including New York correspondent, political reporter, investigative reporter, and foreign correspondent. Harden is the author of *Africa: Dispatches from a Fragile Continent* and *A River Lost: The Life and Death of the Columbia.* He lives in New York City.

Robert M. Hazen is a research scientist at the Carnegie Institution of Washington's Geophysical Laboratory and the Clarence Robinson Professor of Earth Science at George Mason University. He is the author of more than 230 articles and 15 books, including *The Music Men, Wealth Inexhaustible,* and *Keepers of the Flame* (all coauthored by his wife, Margaret Hindle Hazen), as well as *The Breakthrough, The New Alchemists, Why Aren't Black Holes Black?,* and *The Diamond Makers.* A professional trumpeter, Hazen lives with his wife in Glen Echo, Maryland.

Sarah Blaffer Hrdy is an anthropologist who studies primates, including humans. She has done fieldwork in Central America, Madagascar, and India and is the author of *The Black-man of Zinacantan, The Langurs of Abu: Female and Male Strategies of Reproduction, The Woman That Never Evolved,* and, most recently, *Mother Nature: A History of Mothers, Infants, and Natural Selection.* The mother of three children, Hrdy lives with her husband on a farm in northern California, where they are involved in farming and habitat restoration.

Garret Keizer is the author of three books of nonfiction, *No Place But Here, A Dresser of Sycamore Trees,* and, most recently, *The Enigma of Anger: Essays on a Sometimes Deadly Sin,* as well as the novel *God of Beer.* He lives in northeastern Vermont with his wife and daughter.

Verlyn Klinkenborg is the author of *Making Hay, The Last Fine Time,* and *The Rural Life,* to be published in January 2003. He is a member of the *New York Times* editorial board and lives with his wife in rural New York State.

Robert Kunzig writes for *Discover* and other magazines from Dijon, France. His book on oceanography, *Mapping the Deep,* won the 2001 Aventis Prize for the best science book of the year.

Harry Marshall is the head of Icon Films in Bristol, England. He was born and raised in India and has worked extensively in the Himalayas. In 1999 he won the Writers Guild of America Best Writing Award at the Jackson Hole Wildlife Film Festival. In 2000 he directed and produced *Abominable Snowman: The Search for the Truth* for Channel 4 in the United Kingdom and the Learning Channel in the United States.

Anne Matthews is the author of three books about American places undergoing profound change: *Wild Nights: Nature Returns to the City, Where the Buf-*

falo Roam, and *Bright College Years.* Her work has appeared in *Outside, Architecture,* the *Washington Post,* and the *New York Times.* She is a contributing editor at *Preservation* and teaches at Princeton University. She lives in Princeton, New Jersey.

Steve Mirsky is an editor at *Scientific American,* where he has written the "Antigravity" column, a look at the lighter side of science, for the past seven years. He compares doing a humor column at *Scientific American* to making the best sloppy joes at the Culinary Institute. He has written about science for a wide variety of magazines, including *Audubon, Astronomy, Earth,* and *Wildlife Conservation,* and has been a commentator for National Public Radio. Mirsky sneaked out of Cornell University with a master's degree in chemistry.

Judith Newman is a contributing editor for *Allure* and *Self.* She has written about medicine, business, entertainment, and popular culture for a wide variety of publications, including *Vanity Fair, Harper's Magazine, Discover, Newsweek,* the *New York Times,* and *GQ.* Her book, *You Make Me Feel Like an Unnatural Woman: The First Year in the Life of an Old Mother and New Twins,* is forthcoming. She lives with her husband, twin sons, and golden retriever in Manhattan.

Dennis Overbye is a science reporter for the *New York Times.* He is the author of *Einstein in Love: A Scientific Romance,* which was a Los Angeles Times Book Prize finalist, and *Lonely Hearts of the Cosmos: The Story of the Scientific Quest for the Secret of the Universe,* which was a finalist for the National Book Critics Circle Award and the Los Angeles Times Book Prize and won the American Institute of Physics award for science writing. He lives in Manhattan with his wife, Nancy Wartik.

Chet Raymo is professor emeritus at Stonehill College in Easton, Massachusetts. He is author of ten books on science and nature, including *The Soul of the Night, Honey from Stone, Natural Prayers,* and, most recently, *An Intimate Look at the Night Sky.* His weekly column, "Science Musings," has appeared in the *Boston Globe* since 1983. His work has been widely anthologized and is included in *The Norton Book of Nature Writing,* among others. He won the Lannan Literary Award in 1998 for his nonfiction work.

Eric Schlosser is a correspondent for *The Atlantic Monthly* and the author of the best-selling book *Fast Food Nation.* After the publication of "Why McDonald's Fries Taste So Good," McDonald's restaurants were destroyed by radical Hindus, class action lawsuits were filed against the restaurant chain, and riots broke out in India. These actions forced apologies from McDonald's, a corporate acknowledgment that the fries contained beef, and donations to vegetarian and Hindu causes. Schlosser lives in New York City.

Daniel Smith graduated from Brandeis University and now works as a freelance writer in New York City. He is presently writing a book on the history and science of hearing voices.

Peter Stark is a contributor to *Outside, Smithsonian,* and *The New Yorker.* "The Sting of the Assassin" is included in his book *Last Breath: Cautionary Tales from the Limits of Human Endurance.* He has been nominated for a National Magazine Award and has published a collection of essays, *Driving to Greenland.* He is also the editor of an anthology of writing about the Arctic, *Ring of Ice.* He lives in Missoula, Montana.

Clive Thompson writes about technology, culture, and politics. He is an editor at large for *Shift,* a magazine of digital culture, and contributes regularly to such publications as the *New York Times Magazine, Wired, Entertainment Weekly,* the *Washington Post, Elle Canada,* the *Baffler,* the *Toronto Star,* and *Report on Business.* He can be reached at clive@bway.net.

Joy Williams is the author of four novels, *State of Grace, The Changeling, Breaking and Entering,* and *The Quick and the Dead;* two short story collections, *Taking Care* and *Escapes;* and a collection of essays, *Ill Nature,* which was a finalist for the National Book Critics Circle Award. She has received the Rea Award for the short story and the Strauss Living Award from the American Academy of Arts and Letters. *The Quick and the Dead* was a finalist for the Pulitzer Prize in 2000.

Karen Wright writes a monthly column called "Works in Progress" for *Discover.* Her work has also appeared in the *New York Times Magazine,* the *New York Times Book Review,* the *Los Angeles Times, Nature, Science,* and *Scientific American.* She is the recipient of an Evert Clark Award for science journalism and a Knight Science Journalism Fellowship from the Massachusetts Institute of Technology. She would like to thank Steve Petranak, Sarah Richardson, Dave Grogan, and Sara Pratt for their help in producing this article.

Other Notable Science and Nature Writing of 2001

SELECTED BY TIM FOLGER

MARK ALPERT
Kibbles and Bytes. *Scientific American*, June.
A. ALVAREZ
Ice Capades. *New York Review of Books*, August 9.
NATALIE ANGIER
Altruism, Heroism, and Nature's Gifts in the Face of Terror. *New York Times*, September 18.
ANTHONY F. AVENI
Other Stars Than Ours. *Natural History*, April.

RICK BASS
Why I Hunt. *Sierra*, July/August.
SHARON BEGLEY
Assessing the Threat of "Bugs" and "Gas." *Newsweek*, October 8.
The Brain in Winter. *Newsweek*, Fall/Winter special issue.
STEVEN BODIO
Sovereigns of the Sky. *Atlantic Monthly*, February.

RICHARD CONNIFF
Following the Track of the Cat. *Smithsonian*, July.
Monkey Wrench. *Smithsonian*, October.
BERNARD J. CRESPI
Altruism in the Outback. *Natural History*, November.
KEN CROSWELL
Wondering in the Dark. *Sky & Telescope*, December.

JARED DIAMOND
Anatomy of a Ritual. *Natural History*, July/August.

JULIAN DIBBELL
Lost Love. *Harper's Magazine,* August.
BILL DONAHUE
America's Little (Well . . .) Actually Kind of Serious (Um . . .) Maybe It's Worse Than We Thought (Hmmm . . .) Pretty Damn Big (Gulp!) Arsenic Problem. *Outside,* February.
JØRGEN DRAGSDAHL
The Danish Dilemma. *Bulletin of the Atomic Scientists,* September/October.

EVAN EISENBERG
The Adoption Paradox. *Discover,* January.
CAROL EZZELL
The Himba and the Dam. *Scientific American,* June.

CAROLINE FRASER
The Ballad of Lonesome George. *Outside,* January.

W. WAYT GIBBS
Art as a Form of Life. *Scientific American,* April.
On the Termination of Species. *Scientific American,* November.
JANET R. GILSDORF
A Simple Song of Gratitude. *JAMA,* March 28.
MALCOLM GLADWELL
Smaller. *New Yorker,* November 26.
ADAM GOODHEART
Bringing Back the Beast. *Outside,* March.
JEROME GROOPMAN
Eyes Wide Open. *New Yorker,* December 3.
The Thirty Years' War. *New Yorker,* June 4.
DANIEL GROSSMAN
Dissent in the Maelstrom. *Scientific American,* November.
DEREK GRZELEWSKI
Risky Business. *Smithsonian,* November.

JOHN HARLIN
A Narrative of the Savage Coast. *Backpacker,* September.
TOM HILDITCH
The Elephants of Bangkok. *Utne Reader,* December.
JACK HITT
Dinosaur Dreams. *Harper's Magazine,* October.

VERLYN KLINKENBORG
Cow Parts. *Discover,* August.
KEVIN KRAJICK
Message in a Bottle. *Smithsonian,* July.

ROBERT KUNZIG
They Love the Pressure. *Discover,* August.
Trapping Light. *Discover,* April.

BRUCE MCCALL
Rat Dreams. *Discover,* October.
SUSAN MILIUS
Face the Music. *Natural History,* December.
STEVE MIRSKY
A Host with Infectious Ideas. *Scientific American,* May.
Nostrildamus. *Scientific American,* June.
Out of This World. *Scientific American,* June.

SHERWIN NULAND
The Breast Cancer Wars. *New York Review of Books,* September 20.

STEVE OLSON
The Genetic Archaeology of Race. *Atlantic Monthly,* April.
JEREMIAH P. OSTRIKER AND PAUL J. STEINHARDT
The Quintessential Universe. *Scientific American,* January.

JAKE PAGE
Seeing Fingers Decipher Bones. *Smithsonian,* May.
P. JAMES E. PEEBLES
Making Sense of Modern Cosmology. *Scientific American,* January.
DAVID PETERSEN
A Postmodern Awakening. *Orion,* Winter.
HEATHER PRINGLE
The Creature from the Zuni Lagoon. *Discover,* August.
Gladiatrix. *Discover,* December.
Secrets of the Alpaca Mummies. *Discover,* April.

PAUL RAEBURN
Down in the Dirt Wonders Beckon. *Business Week,* December 3.
BILL ROORBACH
Temple Stream. *Harper's Magazine,* December.
ADRIENNE ROSS
Salvage. *EarthLight,* Winter.

SCOTT RUSSELL SANDERS
Stillness. *Orion,* Spring.
ROBERT M. SAPOLSKY
Fossey in the Mist. *Discover,* February.
MARK SINCELL
Twin Stars of Astrophysics Make Room for Two. *Science,* August 10.

NATASHA SINGER
Jonah Is the Whale. *Outside,* August.
FLOYD SKLOOT
Counteracting the Powers of Darkness. *Virginia Quarterly Review,* Autumn.
LAUREN SLATER
Dr. Daedalus. *Harper's Magazine,* July.
ANNICK SMITH
Whale Song. *Audubon,* May/June.
RICHARD STONE
The Cold Zone. *Discover,* February.

GARY TAUBES
The Soft Science of Dietary Fat. *Science,* March 30.
NEIL DE GRAASE TYSON
By Any Other Name. *Natural History,* July/August.
Fear of Numbers. *Natural History,* December.
Over the Rainbow. *Natural History,* September.

STEVEN WEINBERG
The Future of Science, and the Universe. *New York Review of Books,* November 15.
JULIA WHITTY
Shoals of Time. *Harper's Magazine,* January.
KAREN WRIGHT
Splendor in Glass. *Discover,* January.